BASIC/NOT BORING
SOCIAL STUDIES SKILLS

MIDDLE GRADE BOOK OF SCIENCE TESTS

Series Concept & Development
by Imogene Forte & Marjorie Frank

Illustrations by Kathleen Bullock

Incentive Publications, Inc.
Nashville, Tennessee

About the cover:
Bound resist, or tie dye, is the most ancient known method of fabric surface design. The brilliance of the basic tie dye design on this cover reflects the possibilities that emerge from the mastery of basic skills.

Cover art by Mary Patricia Deprez, dba Tye Dye Mary®
Cover design by Marta Drayton and Joe Shibley
Edited by Jean K. Signor

ISBN 0-86530-494-7

Copyright ©2001 by Incentive Publications, Inc., Nashville, TN. All rights reserved. No part of this publication may be reproduced, stored in a retrieval system, or transmitted in any form or by any means (electronic, mechanical, photocopying, recording, or otherwise) without written permission from Incentive Publications, Inc., with the exception below.

Pages labeled with the statement ©2001 by Incentive Publications, Inc., Nashville, TN are intended for reproduction. Permission is hereby granted to the purchaser of one copy of **BASIC/NOT BORING MIDDLE GRADE BOOK OF SCIENCE TESTS** to reproduce these pages in sufficient quantities for meeting the purchaser's own classroom needs only.

PRINTED IN THE UNITED STATES OF AMERICA
www.incentivepublications.com

TABLE OF CONTENTS

Inside the Middle Grade Book of Science Tests 7

How to Use the Middle Grade Book of Science Tests 8

SCIENCE CONCEPTS & PROCESSES 10–29
    Science Concepts & Processes Skills Checklists 10
    Test # 1 Nature & History of Science 12
    Test # 2 Science Concepts 20
    Test # 3 Science Processes 26

LIFE SCIENCE 32–59
    Life Science Skills Checklists 32
    Test # 1 Life Characteristics & Processes 35
    Test # 2 Simple Organisms 40
    Test # 3 Plants 44
    Test # 4 Animals 52
    Test # 5 Ecology 60

HUMAN BODY & HEALTH 66–91
    Human Body & Health Skills Checklists 66
    Test # 1 Body Systems, Part 1 69
    Test # 2 Body Systems, Part 2 76
    Test # 3 Diseases & Disorders 82
    Test # 4 Fitness & Health 86

PHYSICAL SCIENCE 92–121
    Physical Science Checklists 92
    Test # 1 Matter: Structure & Properties 94

Test # 2 Matter: Changes & Interactions	102
Test # 3 Force & Motion	106
Test # 4 Energy	110
Test # 5 Waves, Sound, & Light	118

EARTH & SPACE SCIENCE 122–152

Earth & Space Science Skills Checklists	122
Test # 1 Space Science	125
Test # 2 Earth Composition	130
Test # 3 Earth Changes & Processes	134
Test # 4 Rivers, Streams, Lakes, & Groundwater	140
Test # 5 Oceans	144
Test # 6 Air, Weather, & Climate	148

KEEPING TRACK OF SKILLS 153

Student Progress Record	154
Class Progress Record (Science Concepts & Processes, Life Science)	156
Class Progress Record (Human Body & Health, Physical Science)	157
Class Progress Record (Earth & Space Science)	158
Good Skill Sharpeners for Science	159

ANSWER KEYS 161

Science Concepts & Processes Answer Keys	162
Life Science Answer Keys	164
Human Body & Health Answer Keys	167
Physical Science Answer Keys	169
Earth & Space Science Answer Keys	173

INSIDE THE
MIDDLE GRADE BOOK OF SCIENCE TESTS

"I wish I had a convenient, fast way to assess basic skills and standards."
"If only I had a way to find out what my students already know about science!"
"If only I had a good way to find out what my students have learned!"
"How can I tell if my students are ready for state assessments?"
"It takes too long to create my own tests on the units I teach."
"The tests that come with my textbooks are too long or dull."
"I need tests that cover all the skills on a topic—not just a few here and there."

This is what teachers tell us about their needs for testing materials. If you, too, are looking for quality, convenient materials that will help you gauge how well students are moving along towards mastering basic skills and standards—look no further. This is a book of tests such as you've never seen before! It's everything you've wanted in a group of ready-made science assessments for middle grade students.

- The tests are student-friendly. One glance through the book and you will see why. Students will be surprised that it's a test at all! The pages are inviting and fun. Four inquisitive scientists ramble through the pages, leading students through unique and interesting challenges and investigations. Your students will not groan when you pass out these tests. They'll want to stick with them all the way to the end to see where the STOP sign is this time!

- The tests are serious. Do not be fooled by the catchy characters and visual appeal! These are serious, thorough assessments of basic content. As a part of the BASIC/Not Boring Skills Series, they give broad coverage of skills with a flair that makes them favorites of teachers and kids.

- The tests cover all the basic skill areas for science. There are 23 tests within 5 areas: science concepts and processes, life science, human body and health, physical science, and earth and space science.

- The tests are ready to use. In convenient and manageable sizes, each test covers a skill area (such as body systems or changes in matter) that should be assessed. Just give the pages to an individual student, or make copies for the entire class. Answer keys (included in back) are easy to find and easy to use.

- Skills are clearly identified. You can see exactly which skills are tested by reviewing the list of skills provided with each group of tests.

HOW TO USE THE MIDDLE GRADE BOOK OF SCIENCE TESTS

Each test can be used in many different ways. Here are a few:
- as a pre-test to see what a student knows or can do on a certain science topic
- as a post-test to find out how well students have mastered a content or skill area
- as a review to check up on student mastery of standards or readiness for state assessments
- as a survey to provide direction for your present or future instruction
- as an instructional tool to guide students through a review of a lesson
- with one student in an assessment or tutorial setting
- with a small group of students for assessment or instruction
- with a whole class for end-of-unit assessment

The book provides you with tools for using the tests effectively and keeping track of how students are progressing on skills or standards:

- **23 Tests on the Topics You Need:** These are grouped according to broad topics within science. Each large grouping has three or more sub-tests. Tests are clearly labeled with subject area and specific topic.

- **Skills Checklists Correlated to Test Items:** At the beginning of each group of tests, you'll find a list of the skills covered. (For instance, pages 10 and 11 hold lists of skills for the three tests on science concepts and processes.) Each skill is matched with the exact test items assessing that skill. If a student misses an item on the test, you'll know exactly which skill needs sharpening.

- **Student Progress Records:** Pages 154 & 155 hold a reproducible form that can be used to track individual student achievement on all the tests in this book. Make a copy of this form for each student, and record the student's test scores and areas of instructional need.

- **Class Progress Records:** Pages 156–158 hold reproducible forms for keeping track of a whole class. You can record the dates that tests are given, and keep comments about what you learned from that test as well as notes for further instructional needs.

- **Reference for Skill-Sharpening Activities:** Pages 159–160 describe a program of appealing exercises designed to teach, strengthen, or reinforce basic science skills and content. The skills covered in these books are correlated to national curriculum standards and the standards for many states.

- **Answer Keys:** An easy-to-use answer key is provided for each of the 23 tests (see pages 162–175).

THE MIDDLE GRADE SCIENCE TESTS

Science Concepts & Processes Skills Checklists 10–11

    Science Concepts & Processes Tests 12–31

Life Science Skills Checklists 32–34

    Life Science Tests 35–65

Human Body & Health Skills Checklists 66–68

    Human Body & Health Tests 69–91

Physical Science Skills Checklists 92–93

    Physical Science Tests 94–121

Earth & Space Science Skills Checklists 122–124

    Earth & Space Science Tests 125–152

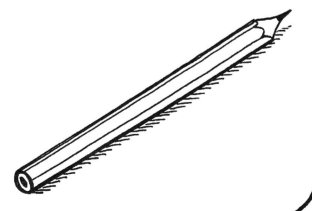

Science Concepts & Processes Skills Checklists

Science Concepts & Processes Test # 1:
NATURE & HISTORY OF SCIENCE

Test Location: pages 12–19

Skill	Test Items
Define science	1
Show understanding of the nature of scientific research and discoveries	2, 3, 11–30
Show understanding of science as a human endeavor	4, 5, 8, 10
Distinguish among different branches of science	6, 7
Identify the topics and areas associated with different branches of science	6, 7, 14–30
Show understanding of the limitations of science	9
Identify pursuits and studies associated with different kinds of scientists	21–30
Identify some uses of science and technology	31–40
Identify discoveries and areas of research associated with specific scientists in history	41–53, 56, 57, 59
Identify some aspects of the history of science	41–60
Show understandings of the relationship between science and technology	61, 62
Show understandings of the relationship between natural disasters and science	63
Show understanding of the benefits and consequences to society of science and technology	64–67
Show understanding of the nature and method of scientific inquiry	68–75

Science Concepts & Processes Test # 2:
SCIENCE CONCEPTS
Location: pages 20–25

Skill   *Test Items*

Show understanding of the concepts of change and constancy 1–7
Show understanding of the concept of cycle 8–9
Show understanding of the concept of cause and effect 10–12
Show understanding of the concept of systems 13–17
Show understanding of how a change in one component changes the system 16–17
Show understanding of the concepts of order and organization 18–21
Show understanding of the concept of equilibrium 22
Show understanding of the concept of evolution 23
Show understanding of the concept of form and function 25–30

Science Concepts & Processes Test # 3:
SCIENCE PROCESSES
Test Location: pages 26–31

Skill   *Test Items*

Understand and use the process of observation 1–4
Understand and use the process of forming a hypothesis 4
Understand and use the process of classification 5–7
Understand and use the process of measurement 8–10
Understand and use the process of using numbers 8–10, 13
Understand and use the processes of identifying and controlling variables 11–12
Understand and use the process of interpreting data 13
Understand and use the process of predicting 14
Understand and use the process of designing an experiment 15
Understand and use the process of using models 16–20
Define terms related to science processes 21–33
Understand and distinguish between
    characteristics of scientific laws and theories 34–36
Understand the process of conducting an experiment 37–39
Recognize some safety procedures for science experimentation 40

Science Concepts, Test #1

NATURE & HISTORY OF SCIENCE

Name __Jamie Cheng__

Date _____

Possible Correct Answers: 75

Your Correct Answers: _____

1. Science comes from the Latin word *scientia*. Most likely, this word means
   a. understanding how the universe works.
   b. knowing or being skilled.
   c. designing experiments.
   d. studying the classifications of life.

Read each group of statements below. Circle the letter of the statement that is true.

2. a. All scientific ideas are subject to change.
    b. Most scientific ideas are proven to be true.
    c. Many scientific ideas are absolutely true; some are subject to change.

3. a. When different scientists research the same problem, they keep working until they all agree on the results before they publish their studies.
    b. Scientists researching the same problem often find and publish different results.
    c. Scientists researching the same problem usually publish different results

4. a. Most scientists have the same set of skills or abilities.
    b. Scientists have a wide range of different interests, skills, and abilities.
    c. All scientists within a particular field of science share the same abilities.

5. a. Most important scientific discoveries have been made by men in the Western Hemisphere.
    b. Most important contributions to science have been made by men in the Western Hemisphere.
    c. Important contributions to science have been made by men and women of many different cultures, countries, and ethnic backgrounds.

How did Homer do on his test? Circle items that he has answered correctly.

TEST: THE NATURE of SCIENCE

Name_____

Directions: Write **T** for True and **F** for False.

__T__ 6. Any branch of science that deals with social or cultural aspects of human behavior or activities is considered to be a social science.

__T__ 7. Physical sciences primarily have to do with studying physical characteristics and processes of living organisms.

__F__ 8. Personal beliefs do not affect the conclusions scientists reach in their research.

__T__ 9. Eventually science and technology will be able to solve all human problems.

__T__ 10. Science requires the use of human qualities such as creativity and honesty.

__T__ 11. All the different sciences have the same goals.

__F__ 12. Fields of science often overlap with one another.

__T__ 13. All scientists spend most of their time doing research.

Which branch of science is concerned with each of these? Write the correct letter.

a. aeronautics
b. genetics
c. meteorology
d. nuclear physics
e. electronics
f. botany
g. archaeology
h. economics
i. zoology
j. physiology
k. seismology

_____ 14. weather

_____ 15. animals

_____ 16. plants

_____ 17. earthquakes

_____ 18. aircraft design and navigation

_____ 19. heredity

_____ 20. materials from former cultures

Who would you (most likely) expect to find studying…?

21. eye movements?
a. an ophthalmologist
b. an engineer
c. a physicist
d. an astronomer

22. insects?
a. an astronomer
b. an etymologist
c. a sociologist
d. a meteorologist

23. human emotions?
a. a psychologist
b. a zoologist
c. a chemist
d. a botanist

24. bones?
a. an anatomist
b. a microbiologist
c. a civil engineer
d. a mechanical engineer

25. earth movements?
a. a geneticist
b. an organic chemist
c. a geologist
d. an immunologist

26. cultures?
a. an aeronautic engineer
b. an anthropologist
c. a rheumatologist
d. a neurologist

27. atoms and molecules?
a. a political scientist
b. a mathematician
c. a psychologist
d. a physicist

28. soil and crop-raising?
a. a sociologist
b. an agronomist
c. a linguist
d. a physiologist

29. blood?
a. a microeconomist
b. an astronomer
c. a geologist
d. a hematologist

30. financial systems?
a. an agronomist
b. an ecologist
c. an economist
d. an embryologist

Describe a way that science is used in each of these locations.

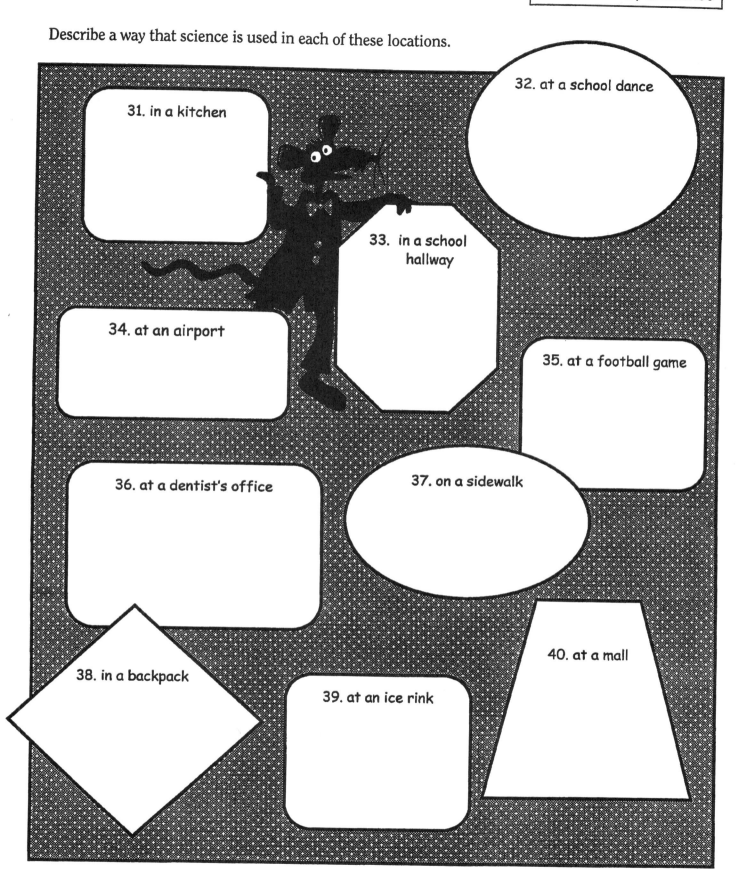

31. in a kitchen
32. at a school dance
33. in a school hallway
34. at an airport
35. at a football game
36. at a dentist's office
37. on a sidewalk
38. in a backpack
39. at an ice rink
40. at a mall

Nature & History of Science

Each scientist's name listed below is connected with research, discoveries, or inventions related to one of the topics on the list.

Write the topic next to the name of each scientist.

41. Johann Gutenberg _____
42. James Watt _____
43. Louis Pasteur _____
44. Copernicus _____
45. Albert Einstein _____
46. Gregor Mendel _____
47. Archimedes _____
48. James Joule _____
49. Hippocrates _____
50. Benjamin Franklin _____
51. Edward Jenner _____
52. Heinrich Hertz _____
53. Marie & Pierre Curie _____

space-time relativity
astronomy
electricity
printing press
smallpox vaccinations
MOTION
eyeglasses
heredity
radium
planetary orbits
mathematics
germs
evolution
laser beams
radio waves
DNA
GRAVITY
steam engine
MEDICINE
heat

54. Which came first?
   a. knowledge of geometry
   b. development of the printing press
   c. study of genetics
   d. system of animal classification

55. Which is most recent?
   a. discovery of radioactivity
   b. nuclear fission
   c. use of radio waves
   d. human trials of gene therapy

56. This physicist helped to develop the atomic bomb. He later expressed concern about the uses of nuclear weapons, particularly the hydrogen bomb. Who was he?
   a. Albert Einstein
   b. Alfred Nobel
   c. Robert Oppenheimer
   d. Lord Kelvin

57. In the mid-1800s, Dmitri Mendelev developed the Periodic Table. Which branch of science was furthered by this development?
   a. genetics
   b. chemistry
   c. astronomy
   d. psychology

58. The ancient Romans were known for amazing developments in the practical science of
   a. building railroads.
   b. building roads and aqueducts.
   c. building pyramids.
   d. developing navigation systems.

59. The Ptolemaic Theory was questioned by Copernicus. This theory argued that
   a. the Earth was the center of the universe.
   b. the sun was the center of the universe.
   c. the Earth rotated on an axis.
   d. the Earth revolved around other planets.

60. The discoveries of Isaac Newton had to do with
   a. biochemistry.
   b. medicine.
   c. nuclear energy.
   d. motion and force.
   e. plate tectonics.

| Nature & History of Science |

61. Which statements are accurate in describing a relationship between science and technology? (*Circle one or more answers.*)
    A. Science explores questions about the world that require instruments or tools to answer.
    B. Scientists may identify problems that a technological tool can solve.
    C. New technology, such as cellular phones, may cause problems that scientists explore to find explanations or to suggest remedies.

62. Which statements are accurate in describing a difference between science and technology? (*Circle one or more answers.*)
    A. Science proposes answers for things happening in the natural world. Technology proposes solutions for human needs.
    B. Scientific study has consequences. The development of a new technology does not.
    C. The risks of scientific study are much greater than the risks of technology.

63. Name three natural hazards that scientists cannot control, but can develop some ways to lower the damage to human and animal life and property.

    A._____ B._____ C._____

64–67: Describe a benefit and a consequence to society for each technological advance:

64. The Internet
Benefit:_____
Consequence:_____

65. Jet Airplanes
Benefit:_____
Consequence:_____

66. Cellular Phones
Benefit:_____
Consequence:_____

67. Automated Saws in Lumber Mills
Benefit:_____
Consequence:_____

Name_____

While playing a game of tennis one day, Andy and Angie noticed that some of the balls had more bounce than others. All the balls were the same kind, made by the same company. They wondered about the difference in the bounce. When they talked about where the balls had come from, they discovered that some of them came from newly-opened cans, and others had been lying around for a while. They guessed that the newer balls were the ones with more bounce. So, they decided to do an experiment to find out if that might be the explanation for the differences.

Andy gathered some balls that he had opened a year ago. Angie bought 8 new cans of tennis balls. On the first day of their experiment, they drew a measuring guide on the wall of Andy's garage. They bounced 6 of the year-old balls, and recorded the height of the bounce. Then they wrote the average of the 6 bounces. Next, they opened 2 new cans (3 balls to a can). They bounced 6 brand new balls, recorded the heights of the bounces, and wrote an average of the 6 bounces.

They opened the other 6 cans of balls and separated them into three groups of 6 balls each. After 3 months, they bounced 6 balls and recorded the measurements in the same way as before. After 6 months, they tested 6 more. After 9 months, they tested the remaining 6.

Angie and Andy looked at the measurements that they had taken over 9 months, as well as the measurements of the new and year-old balls. They put the information onto a bar graph. They came to the conclusion that the longer a tennis ball is out of the can, the less bounce it will have.

68. What observation led the players to do an experiment? _____

69. What question did they want to answer? _____

70. What was their hypothesis? _____

71. What kind of data did they collect? _____

72. How did they collect the data? (How often? In what way?) _____

73. How did they organize their results? _____

74. How did they present their results? _____

75. How did they interpret their data? _____

Science Concepts & Processes Test #2

SCIENCE CONCEPTS

Name _____

Date _____

Possible Correct Answers: 30

Your Correct Answers: _____

For questions 1–5, refer to the list of changes.
Write one or more letters to answer each question.

A. Sam pushes his little brother on a swing.
B. Amy's ice cream bar melts in the sun.
C. Water in the cat's dish evaporates.
D. A campfire burns up ten logs.
E. Ice cream and milk are blended into a milkshake.
F. A piece of toast burns in the toaster.
G. Left outside in the rain, a hammer rusts.
H. A mixture of baking soda and vinegar produces fizz.

_____ 1. Which cause changes in chemical properties of matter?

_____ 2. Which are changes in position of matter?

_____ 3. Which cause only physical changes in the matter?

_____ 4. Which changes involve interactions between substances?

_____ 5. In which example is energy transferred?

6. Which of these are constant?
   (Circle one or more answers.)
   a. the speed of light
   b. the number of cells in the human body
   c. the number of stars in the universe
   d. the charge of an electron

7. Which of these are true?
   (Circle one or more answers.)
   a. Energy can be created.
   b. Energy can change form.
   c. Energy can be destroyed.
   d. The sum of the amount of energy and matter in the universe is constant.

8. Which of the following is NOT an example of the concept of CYCLE?
   a. predator-prey populations
   b. the drift of continents
   c. plant growth and death
   d. the appearance of comets
   e. the ocean tides
   f. the planets orbiting the sun

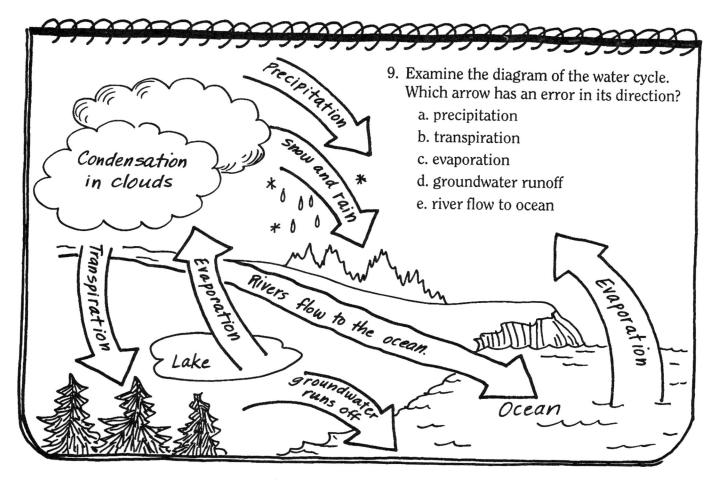

9. Examine the diagram of the water cycle. Which arrow has an error in its direction?
   a. precipitation
   b. transpiration
   c. evaporation
   d. groundwater runoff
   e. river flow to ocean

In each of the following pairs of statements, one states a cause and the other states an effect. Write C for CAUSE or E for EFFECT beside each statement.

10. ____ A. Forests have been cut down and industrialization has increased.
    ____ B. The amount of oxygen provided by plants no longer equals the amount consumed by animals.

11. ____ A. The core of an ordinary star collapses.
    ____ B. The star becomes a white dwarf.

12. ____ A. Air is expelled from the lungs of a gorilla.
    ____ B. The gorilla's diaphragm pushes up and the ribs move in to decrease her chest space.

13. Earth's solar system is a part of some larger systems. Which are examples of these?
   a. the Earth-moon system
   b. the Jupiter-moon system
   c. the Milky Way Galaxy
   d. the universe
   e. the galaxy supercluster

14. Which of these are components of the body's circulatory system?
   (Circle one or more answers.)
   a. spinal cord
   b. neurons
   c. pituitary gland
   d. capillaries
   e. heart
   f. plasma
   g. alveoli
   h. arteries

15. Which of these is probably NOT a component of the ecosystem of a woodland pond? (Circle one or more answers.)
   a. snails and slugs
   b. moss
   c. jellyfish
   d. algae
   e. frogs

16. In this food chain, what will happen if a disease suddenly caused a drastic drop in the number of owls?

_____
_____

17. In this food chain, what would happen if there was a drastic drop in the number of acorns?

_____
_____

18. This is a list of components of life systems. Number the items in the order of complexity. Give a 1 to the most complex component and a 6 to the simplest component.

   _____ populations
   _____ organisms
   _____ cells
   _____ organs
   _____ tissues
   _____ communities

19. Number these events in the order they occur, beginning with winter solstice.

   _____ winter solstice
   _____ summer solstice
   _____ spring equinox
   _____ fall equinox

20. Number the components of animal classification in order of complexity, with 1 being the most complex and 7 being the simplest.

   _____ phylum
   _____ species
   _____ class
   _____ genus
   _____ order
   _____ kingdom
   _____ family

21. The Arctic tern makes the longest migration of any animal. During the Arctic winter, the species travels from their breeding ground in the Arctic to Antarctica. They travel back to the Arctic as the weather turns, making a round trip that totals 22,000 miles a year. This pattern is an example of
   a. energy and matter.
   b. probability.
   c. force.
   d. order and organization.
   e. equilibrium.

Science Concepts

22. Which examples below demonstrate the concept of equilibrium? _____

23. Which examples below demonstrate the concept of evolution? _____

A.
1. A nebula breaks up into a cluster of baby stars.
2. One of these stars shines for pillions of years by turning hydrogen into helium.
3. The star's core collapses under the force of gravity, and the star burns hotter.
4. With greater heat, the helium causes a reaction that blows up part of the star to form a red giant.
5. The core of the star runs out of fuel and collapses.
6. The star shrinks to a smaller body, called a white dwarf.
7. Eventually the star cools to a dead, black dwarf.

B. A river flows to the ocean, carrying rocks and sediment along with it. Over the years, tons of material is deposited at the river's mouth. This builds up to block the flow of the river. The river forges new paths through the sediment, spreading out in finger-like streams to reach the ocean. A delta has been formed.

C. A person steps into a boat and walks toward the front. With each step forward, the boat slides backwards against the water.

D. Body cells are low on nutrients. A hunger signal alerts the body. The body moves nutrimental materials into the cell.

E. A man exercises vigorously. His muscles use oxygen faster than usual. Sensors tell his body that the carbon dioxide level is rising. This causes him to breathe faster to get more oxygen.

F. A thermostat is set at 75°. The room temperature falls to 70°. A sensor tells the thermostat to turn the heater on. The room reaches 75° again. The thermostat turns the heater off.

G. A girl exercises. Her body becomes warm and begins to sweat. Evaporation of the sweat cools her body, and her body temperature falls. Blood vessels in her skin contract to reduce the heat loss.

Science Concepts

The form of an object, organ, or organism contributes to its function.
For each example below, describe how the form affects its function or use.
(What happens, or what is the object able to DO, because of its form?)

FORM — FUNCTION

24. A liquid is made of molecules that move around more than the molecules in a solid.

25. When racing, a swimmer wears a tight-fitting cap and suit of material designed to reduce friction.

26. A submarine has tanks that can take in or expel air and take in or expel water.

27. A space suit has layers that circulate a cooling liquid.

28. X-Rays have a short wavelength and very high frequency that enables them to pass through most living tissue.

29. Ducks have webbed feet.

30. Jet engines have a compressor that sucks in air and a combustion engine that burns fuel and shoots exhaust gasses out the back of the engine.

Name _____

Science Concepts & Processes Test #3

page 1 of 6

SCIENCE PROCESSES

Name _____

Date _____

Possible Correct Answers: 40

Your Correct Answers: _____

To answer questions 1–4, closely observe the garden pictured below.

1. What observation can you make about the flowers in the east end of the garden?
   _____

2. What observation can you make about the flowers on the northwest side of the garden?
   _____

3. What observation can you make about the flowers on the southwest side of the garden?
   _____

4. State a hypothesis that could explain the differences you observe.
   _____

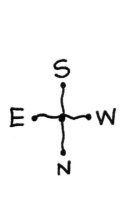

26

Middle Grade Book of Science Tests

Copyright ©2001 by Incentive Publications, Inc., Nashville, TN.

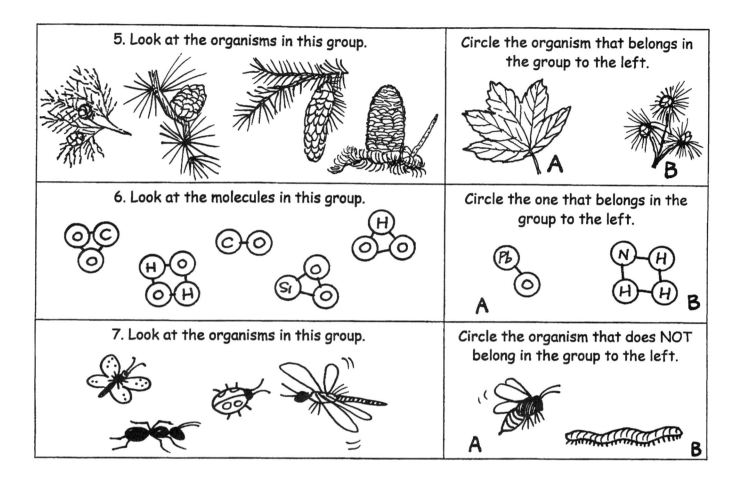

Use the plant diagram below to answer questions 8–10.
Use a centimeter ruler for your measurements.

8. The plant grew about 30 centimeters between Week _____ and Week _____.

9. What was the total growth shown over the 4 weeks? _____

10. How much growth occurred between weeks 2 and 4? _____

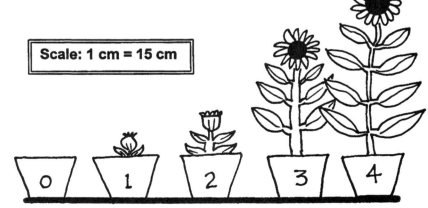

Scale: 1 cm = 15 cm

Science Processes

In Miss Parsnip's science class one day, a rowdy student grabbed a paperweight and a marker off a desk and dropped them out the second story window. He laughed as he watched the objects fall. Much to his surprise, both the objects seemed to hit the ground at the same time.

His friends didn't believe him. They thought the paperweight would certainly fall much faster than the marker. So they decided to do an experiment.

They found two large spoons of exactly the same size and shape. One was made of wood, and weighed 75 grams. The other was made of metal, and weighed 350 grams. They dropped each spoon 5 times, using a stopwatch to time its fall. They calculated an average time for each spoon. In the end, the two spoons had an average fall time that was just hundredths of a second apart.

11. What were the uncontrolled variables?
(Circle one or more.)
a. spoon size
b. spoon weight
c. spoon material
d. distance of the fall
e. landing surface

12. Which variables were controlled?
(Circle one or more.)
a. spoon size
b. spoon weight
c. spoon material
d. distance of fall
e. speed of fall

13. How would you interpret the results of the experiment? _____

14. If the students also timed the fall of a heavy plastic spoon (of exactly the same size and shape) weighing 200 grams, what results do you predict that they would find?

Name _____

Middle Grade Book of Science Tests

At Max's house, four different kinds of bread were kept in a basket for family snacks and meals. Over a few weeks' time, Max noticed that one kind of bread seemed to mold faster than the other kinds. He wanted to find out how fast each kind got moldy, so he could examine the ingredients in the breads and see if he could figure out what kept some bread fresh longer.

15. When Max designs his experiment, which factors are important for him to include in his plan? (*Circle one or more.*)

   a. Start with four kinds of fresh bread, all baked the same day.
   b. Store all kinds of bread in the same kind of a container.
   c. Look at the bread everyday and record observations.
   d. Keep all kinds of bread stored at the same temperature.
   e. Make sure each piece of bread is exactly the same size.
   f. Keep notes about the taste of each kind of bread.

This is a model of one molecule of an element. Use the model for questions 16–20.
(*Note: all neutrons, protons, and electrons are showing in the drawing. None are hidden from the eye.*)

16. How many electrons does this element have? _____

17. How many protons does this element have? _____

18. How many neutrons does this element have? _____

19. How many electrons are in the second level? _____

20. What is the atomic number of this element? _____

Science Processes

Match the science terms with their descriptions. Write a letter for each answer.

A. observation
B. law
C. theory
D. scientific record
E. model
F. problem
G. data
H. hypothesis
I. control
J. inference
K. conclusion
L. mass
M. variable
N. procedure
O. experiment

____21. the amount of matter in an object

____22. a series of steps used to test a hypothesis under carefully controlled conditions

____23. a suggested solution to a problem

____24. a logical answer to a problem based on observations

____25. something that can bring about changes in an experiment

____26. an explanation of things or events (based on many observations)

____27. an example of what might be seen if something could be observed

____28. the way an experiment is carried out

____29. a standard for comparison in an experiment

____30. the meaning drawn from a conclusion

____31. recorded facts or measurements gathered from an experiment

____32. the things noticed by your senses

____33. a statement that describes the way nature works (based on many observations)

Name _____ 30

34. Which is an example of a scientific law?
   a. Energy is neither created nor destroyed.
   b. Dinosaurs disappeared from Earth because of a drastic climate change.
   c. The universe began 15 billion years ago in a huge explosion called "The Big Bang."

35. Which is an example of a scientific theory?
   a. Ice melts when it is heated.
   b. On Earth, an object that is dropped falls to the ground.
   c. The surface of Earth is made of rigid plates that move.

36. What would cause a scientific law or theory to be dropped?

   _____
   _____

37. Why is it important for scientists to repeat experiments?

   _____
   _____

38. Why is it important to have a control in an experiment?

   _____
   _____

39. Why should a scientist test only one variable at a time?

   _____
   _____

SCIENCE SAFETY

A. Get permission before doing an experiment.
B. Keep safety goggles nearby in case you need them.
C. Never eat or drink in a science lab.
D. Wash off any spilled chemicals immediately.
E. Always slant test tubes towards you so you can see what's going on in them.
F. Do not touch chemicals with your bare hands.
G. Only taste substances that you think are safe.
H. Keep a fire extinguisher and first aid kit nearby.
I. If your clothing catches on fire, run out of the room.

40. A science student wrote up some rules for his science lab. Read them over. Cross out any that he has written totally or partially wrong.

Life Science Skills Checklists

Life Science Test # 1:
LIFE CHARACTERISTICS & PROCESSES
Test Location: pages 35–39

Skill	Test Items
Identify characteristics and make-up of living things	1–3
Identify examples of adaptation of living things	4
Identify ways living organisms use energy	5–6
Recognize characteristics and functions of cells and cell processes	7–15
Distinguish between plant and animal cells	16–17
Identify components of cells and their functions	18–21
Identify various cell and life processes	22–32
Show understanding of the system of life classification	33–40
Recognize the make-up of an organism's scientific name	36
Recognize characteristics of organisms in specific life kingdoms	37–40

Life Science Test # 2:
SIMPLE ORGANISMS
Location: pages 40–43

Skill	Test Items
Identify characteristics of organisms in the moneran, protist, and fungi kingdoms; distinguish among the three kinds of organisms	1–7
Show understanding of the definition and characteristics of viruses	8–15
Identify specific characteristics of monerans	16
Identify characteristics and functions of cyanobacteria	17
Define parasites, saprophytes, and mutualism	18–20
Recognize characteristics and functions of bacteria	21–28
Distinguish among different kinds of protists	29–33
Describe functions of cilia, pseudopods, and flagella in protists	34–36
Identify characteristics and functions of protists	37–38
Identify characteristics and functions of fungi	39–45
Distinguish between different kinds of fungi	41–42
Describe ways fungi can be helpful	43–45

Life Science Test # 3:
PLANTS
Location: pages 44–49

Skill	Test Items
Identify characteristics of plants	1
Identify the major divisions of the plant kingdom	2
Distinguish between vascular and nonvascular plants	3–5, 42
Identify the names and functions of plant structures	6–17
Show understanding of key plant processes of photosynthesis, transpiration, respiration, and gas exchange	18–21
Recognize and describe different types of plant behavior	22–30
Identify characteristics of different seed plants	31–35
Show understanding of the system of plant classification	31–40
Identify characteristics of different seedless plants	36–40
Distinguish between seed plants and seedless plants	42
Distinguish between gymnosperms and angiosperms	43–44
Identify names and functions of flower parts	45–48
Show understanding of concepts related to plant reproduction, growth, and development	49–56
Recognize characteristics of different kinds of soils	57–58
Show understanding of the carbon dioxide-oxygen cycle and the nitrogen cycle as related to plant processes	59–60

Life Science Test #4:
ANIMALS
Test Location: pages 50–59

Skill	Test Items
Show understanding of the system of classification of animals	1–9
Recognize characteristics of organisms in the nine major animal phyla	1–9
Recognize examples of organisms belonging to different animal phyla	10–28
Distinguish between bilateral symmetry, radial symmetry, and no symmetry	29–30
Recognize the classifications of vertebrates and invertebrates in the animal kingdom	31
Show understanding of the numbers and complexity of animals in various phyla	32–34
Define the process of molting	35
Recognize characteristics of different classes of arthropods	36–40
Identify animals in different classes of arthropods	41–45
Recognize the stages in complete and incomplete metamorphosis of insects	46–52
Identify characteristics of animals in the different classes of the chordate phylum	53–72
Distinguish among amphibians, reptiles, fish, mammals, and birds	53–72
Show understanding of the reproductive processes of regeneration, budding, external fertilization, and internal fertilization	73–76
Recognize and identify animal behaviors	77–83
Distinguish between acquired and inborn behavior	79–81
Show understandings of concepts related to genetics and heredity	84–90

Life Science Test # 5:
ECOLOGY
Test Location: pages 60–65

Skill	Test Items
Recognize concepts and terms related to ecology	1–6
Recognize and define different relationships within an ecosystem	7–11
Distinguish between producers and consumers	12–17
Distinguish between producers, primary consumers, and secondary consumers in a food chain	14–17
Recognize characteristics of different biomes	18–21
Understand factors that influence the size of a population	22
Understand the concept of succession in a community	23
Identify fossil fuels, biodegradable substances, and nonrenewable resources	24–25
Show understanding of the nitrogen, water, and carbon dioxide-oxygen cycles	27–29
Describe benefits and consequences of technological choices in relation to the environment	30–31
Show understanding of effects of pollutants	32–33
Recognize the value of techniques to conserve resources and protect the environment	34–35

Life Sciences, Test #1

LIFE CHARACTERISTICS & PROCESSES

Name _____

Date _____

Possible Correct Answers: 40

Your Correct Answers: _____

"Attention, students...."

1. Which of these characteristics is not necessarily a characteristic of living organisms only?
   A. It takes in food or substances to make food.
   B. It has a life span: it grows, develops, and dies.
   C. It gives off wastes produced by the cells.
   D. It is made up of matter.
   E. It needs water to survive.
   F. It reproduces itself.
   G. It is made up of one or more cells.
   H. It responds to its surroundings.
   J. It uses energy for its activities.

2. The three basic elements that make up living substances are
   a. oxygen, hydrogen, and nitrogen
   b. air, water, and blood
   c. water, muscles, and vessels
   d. oxygen, hydrogen, and carbon
   e. sodium, hydrogen, and oxygen
   f. carbon, copper, and hydrogen

3. Which statement below is correct? _____

   A. Water makes up 99% of most living things.
   B. Water makes up 50% of most living things.
   C. Water makes up 70% of most living things.

Life Characteristics & Processes

> Fish have gills that get oxygen from water.
>
> With the help of camouflage, some insects blend right in with their leafy surroundings.
>
> A rattlesnake has a noisy rattle to frighten away predators.
>
> Plants have roots that anchor them to the soil.
>
> Earthworms take oxygen into their bodies through their moist skin.

4. The descriptions above are all examples of
   a. reproduction
   b. instinct
   c. adaptation
   d. fertilization
   e. active transport

5. Name two ways a plant uses energy:

_____

_____

6. Name two ways an animal uses energy:

_____

_____

Write T (true) or F (false) for each statement.

____ 7. All organisms are made of many cells.

____ 8. All cells contain a cell membrane and a cell wall.

____ 9. All cells are produced from other cells.

____ 10. The chromosomes carry a code in each cell.

____ 11. The cell membrane controls movement of materials in and out of the nucleus.

____ 12. All cells have a similar function in the organism.

____ 13. The cell's cytoplasm contains the cell materials.

____ 14. Some cells can be seen without a microscope.

____ 15. A molecule is the basic unit of every living organism.

16. Which cell below is a plant cell? _____

17. Which cell below is an animal cell? _____

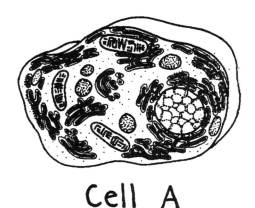

Cell A

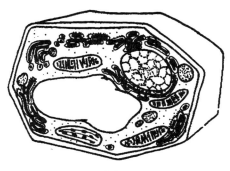

Cell B

18. In a cell, the **chloroplast**
   a. surrounds the plant cell and gives it shape.
   b. stores and releases chemicals for cell use.
   c. controls chlorophyll to help the cell trap light to make food.
   d. holds the code that controls the cell.

19. In a cell, the **mitochondria**
   a. store water and dissolved materials.
   b. control the movement of materials in and out of the cell.
   c. make proteins.
   d. release energy for cell use.

20. Which controls the cell's activities?
   a. cytoplasm
   b. nucleus
   c. cell wall
   d. cell membrane

21. Which is a tube network in cytoplasm where the cell substances are made?
   a. endoplasmic reticulum
   b. ribosomes
   c. Golgi bodies
   d. mitochondria

Life Characteristics & Processes

Choose one of the words on the right to match each description.

WHAT'S GOING ON?

_____ 22. The cells in the root of a bean plant drink in water through the cell membranes.

_____ 23. The nucleus of a skin cell divides into 2 nuclei so 2 new skin cells can form.

_____ 24. Particles of water that are crowded in their space move to a place where the particles are less concentrated.

_____ 25. Cell division produces offspring from only one parent cell.

_____ 26. An organism is growing, repairing, and using food.

_____ 27. A male and female gamete combine.

_____ 28. Thirsty cells in a sunflower have been without water for a long time. The cytoplasm begins to shrink, causing the plant to become limp.

_____ 29. A signal in an organism alerts cells to increase, decrease, or maintain an activity.

_____ 30. Material from the nuclei of two different cells combine to produce offspring.

_____ 31. Cells in a frog break down food in a chemical reaction that releases energy for jumping.

_____ 32. During a long run, a giraffe's body heats up. Soon, the body begins to sweat and to cool off so its temperature doesn't get too high.

asexual reproduction
mitosis
diffusion
homeostasis
FERTILIZATION
OSMOSIS
plasmolysis
feedback
RESPIRATION
metabolism
sexual reproduction

33. The science of classification of living things is
    a. biology   b. taxonomy   c. cytology   d. ecology

34. Name the five kingdoms of living things.

    _____
    _____
    _____
    _____
    _____

35. These are the divisions for classifying living organisms. Number them in order from largest (1) to smallest (7).

    ____ order
    ____ phylum
    ____ kingdom
    ____ species
    ____ class
    ____ family
    ____ genus

36. The scientific name of a lion is **Panthera leo**. Two divisions of classification are used to form the scientific name of each organism.

    Label the divisions: **Panthera** is the _____; **leo** is the _____.

37–40. Write the name of one or more kingdoms to answer each question.

37. I cannot make my own food. I reproduce by forming spores. To which kingdom or kingdoms could I belong?
    _____

38. All organisms in my kingdom make their own food. To which kingdom or kingdoms could I belong?
    _____

39. Most organisms in my kingdom can move. We must find food and water to stay alive. To which kingdom or kingdoms could I belong?
    _____

40. I have only one cell. To which kingdom or kingdoms could I belong?
    _____

SIMPLE ORGANISMS

Professor Ozzie Moses is examining some simple organisms under the microscope.

1. The first one is a one-celled organism. Which of these could it be?
   a. moneran   b. protist   c. fungus

2. The next organism contains chlorophyll. Which of these could it be?
   a. moneran   b. protist   c. fungus

3. The next organism moves with flagella. Which of these could it be?
   a. moneran   b. protist   c. fungus

4. The next organism produces spores. Which of these could it be?
   a. moneran   b. protist   c. fungus

Identify each of these organisms as a moneran, protist, or fungus.

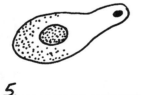
5. _____

6. _____

7. _____

What About Viruses. Answer T (true) or F (false) for each statement.

_____ 8. A virus is a plant.
_____ 9. A virus is an animal.
_____ 10. A virus can be destroyed.
_____ 11. A virus is a living organism.
_____ 12. A virus has properties of both living and nonliving things.
_____ 13. Viruses contain nuclei and other cell parts.
_____ 14. Some viruses can be seen with the naked eye.
_____ 15. Viruses cannot grow outside a living organism.

Simple Organisms

16. An organism in the moneran kingdom (*Circle one or more answers.*)
    a. has one cell and one nucleus.
    b. has many cells and no nuclei.
    c. has one cell with no nucleus.
    d. is classified as a plant.
    e. has no walls around the cells.
    f. is classified as an animal.

17. Cyanobacteria are valuable to other organisms because they
    a. destroy harmful bacteria.
    b. are a food source to other organisms.
    c. are able to take nitrogen from the air and change it into a form that other organisms can use.

Draw a line to the definition that matches each term.

18. parasites   A. a relationship between two organisms where both benefit

19. saprophytes   B. an organism that gets food from dead organisms or waste products

20. mutualism   C. an organism that gets food and protection from a host organism, and often harms the host

Bacteria Quiz

Circle the numbers of true statements.

21. Cocci bacteria are spherical.
22. Bacilli bacteria are rod-shaped.
23. All bacteria need oxygen to live.
24. Bacteria are present in your nose.
25. All bacteria are harmful.
26. Bacteria reproduce by fission.

27. What is the purpose of pasteurization? _____

28. Describe a use or function of helpful bacteria. _____

Simple Organisms

  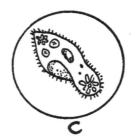 

A   B   C   D

Write the correct letter for each answer.

_____ 29. Which organism under the microscope is a dinoflagellate?

_____ 30. Which organism under the microscope is a euglenoid?

_____ 31. Which organism under the microscope is a paramecium?

_____ 32. Which organism under the microscope is an amoeba?

33. A one-celled, animal-like protist that lives in soil or water or on decaying organic matter is a(n)
_____

34. What is the function of the pseudopods that form in an amoeba?
_____

35. What is the function of the cilia that cover a paramecium?
_____

36. How does a flagellate use its flagella?
_____

37. Most species of plant-like protists live in plankton in lakes and oceans. What benefit do these protists have for animals in the plankton?
_____

38. Malaria is a disease caused by a sporozoan. Why does malaria spread only with the help of both mosquitoes and humans?
   a. The mosquito cannot travel far enough to carry the disease alone.
   b. A sporozoan needs two hosts to complete its life cycle.
   c. The mosquito cannot hold enough blood for the sporozoan to reproduce.

Simple Organisms

39. There are three main phyla in the fungi kingdom. How are the different groupings classified?
    a. by their shape
    b. by the way spores are formed for reproduction
    c. by the environment in which they reproduce

40. Yeasts sometimes reproduce when a new cell grows out of the side of an old cell. This process is called   a. fermentation   b. budding   c. anaerobic respiration

41. Dr. Moses is examining the mold on a piece of bread. What kind of fungus is he probably seeing?
    a. sporangium fungus   b. club fungus   c. sac fungus

42. Dr. Moses picks some mushrooms and puffballs. What kind of fungus is he gathering?
    a. sporangium fungus   b. sac fungus   c. club fungus

43. Mushrooms can be harmful. Some are poisonous to eat. Describe a way in which mushrooms can be **helpful**.
    _____
    _____

44. Yeasts can be harmful. Some can cause infections in the lungs of animals. Describe an example of **helpful** yeast.
    _____
    _____

45. Molds can be harmful. Some molds cause diseases such as athlete's foot. Other molds cause food to spoil. Describe an example of a **helpful** mold.
    _____
    _____

Life Sciences, Test #3

PLANTS

Name _____   Possible Correct Answers: 60

Date _____   Your Correct Answers: _____

1. Circle the statement that is NOT true about plants.
   - Most of us are complex organisms with many cells.
   - We contain chlorophyll.
   - Our cells have walls.
   - We can change locations.
   - We need water to survive.
   - Most of us have roots, stems, and leaves.

2. Which are the five major divisions of the plant kingdom?
   a. green algae, brown algae, red algae, liverworts & mosses, vascular plants
   b. vascular plants, nonvascular plants, seed plants, conifers, and ferns
   c. algae, ferns, mosses, fungi, flowering plants
   d. seed plants, conifers, flowering plants, gymnosperms, angiosperms

3. Which are three examples of nonvascular plants?
   a. ferns, mosses, horsetails
   b. algae, mosses, gymnosperms
   c. algae, liverworts, mosses
   d. algae, viruses, bacteria

4. What is the difference between vascular and nonvascular plants? _____

5. How do plants with no roots or stems take in water? _____

Middle Grade Book of Science Tests   Copyright ©2001 by Incentive Publications, Inc., Nashville, TN.

Plants

Finish each description by writing the correct plant feature.

Word bank: xylem, phloem, cambium, stems, annual, root hairs, biennial, epidermis, evergreen, deciduous, taproot, fibrous root, perennial, herbaceous

6. Roots absorb nutrients through _____, which are threadlike cells that grow near the tip of the root.

7. _____ is a plant tissue made of vessels that carry nutrients around the plant.

8. Some plants have a _____ system, where food is stored in a long, thick main root.

9. _____ trees lose their leaves. _____ trees remain green all year.

10. A(n) _____ plant lives from one season to another.

11. A(n) _____ plant produces leaves and food in one year, then reproduces and dies in the second year.

12. _____ support leaves and allow movement of food and water through the plant.

13. The _____ is a layer of brick-like cells that covers the leaf and protects its inner parts.

14. Trees and grasses have _____ systems; the roots have many branches to hold the soil in place.

15. _____ is plant tissue with tube-like cells that transport food from the leaves to other parts of the plant.

16. A(n) _____ plant grows, reproduces, and dies in one year.

17. _____ is a growth tissue that makes new xylem and phloem cells.

Professor Ozzie Moses is teaching students about plant processes.
He has written statements that help to explain each process.
Some of the statements are wrong.
He wants to see if the students are sharp enough to catch these.
Circle one or more incorrect statements on each poster.

18. GAS EXCHANGE

A. During gas exchange, oxygen, carbon dioxide, and water vapor exit and enter a leaf through stomata.

B. Gases move in and out of the stomata by diffusion.

C. The stomata of a leaf are usually open at night.

D. Guard cells on either side of the stomata control the size of the opening.

E. Each leaf has about 10-100 stomata.

19. PHOTOSYNTHESIS

A. Green plants trap light energy in chlorophyll molecules.

B. Light energy is needed to give energy for the chemical reactions in photosynthesis.

C. In photosynthesis, the plant produces carbon dioxide and water.

D. In photosynthesis, the plant produces oxygen.

E. In photosynthesis, sugar is produced and stored in the plant.

20. RESPIRATION

A. Respiration releases energy for the plant's activities.

B. During respiration, the plant produces sugar.

C. Heat is released from the plant during respiration.

D. Respiration takes place in all the plant's cells.

E. During respiration, the plant releases carbon dioxide and sugar.

F. The plant uses sugar for the chemical reaction of respiration.

21. TRANSPIRATION

A. Transpiration occurs only at night in plants.

B. Transpiration is the opposite process of photosynthesis.

C. During transpiration, the plant loses water vapor through its leaves.

D. Water vapor leaves the plant through the guard cells.

E. Water vapor leaves the plant through the stomata.

F. Water vapor leaves the plant through the process of osmosis.

Each statement describes something that relates to plant behavior. Write the correct word for each one.

PLANT BEHAVIOR
stimulus
response
tropism
phototropism
positive gravitropism
negative gravitropism

_____22. Gravity pulls on plant roots.

_____23. The stem grows upward against the pull of gravity.

_____24. A plant responds to a stimulus.

_____25. Someone touches a plant's leaf.

_____26. Behavior is changed as a result of a stimulus.

_____27. Bright sunlight shines on a plant.

_____28. A plant grows toward the sun.

_____29. Roots grow downward.

30. Which describes a plant response that would be controlled by an auxin?

   a. A peach tree loses ten quarts of water through transpiration on a hot, sunny day.

   b. A beet plant obtains energy from the sugar that was made during photosynthesis.

   c. An birch tree has bark that has a white, papery appearance

   d. Cells on the shaded side of a stem cause the stem cells to grow faster than cells on the lighted side, allowing the stem to curve toward the light.

Plants — page 5 of 8

Labels have fallen off the Plant Classification Chart. Write them back in the correct places.

Plant Classification

Word bank (scattered labels):
- Ferns
- Monocots
- Gymnosperms
- Algae
- SEED PLANTS
- Angiosperms
- Dicots
- Club Mosses & Horsetails
- Liverworts & Mosses
- SEEDLESS PLANTS

31. _____

32. _____
- plants that produce seeds not protected by fruit
  (Examples: conifers such as pine, spruce, fir)

33. _____
- plants that produce seeds inside a fruit
  (Examples: beans, onions, apples)

34. _____
- flowering plants with flower parts in threes • food stored in one seed leaf • vascular bundles scattered throughout the stem

35. _____
- flowering plants with flower parts in fours or fives • food stored in two seed leaves • vascular bundles in a ring inside the stem

36. _____

37. _____
- simple plants with no roots, stems, or leaves
- three kinds: green, red, and brown
- single or many-celled; live near water
- reproduce asexually or by conjugation

38. _____
- simple, nonvascular plants that live and grow in moist and shady places
- rootless plants that attach to the ground by rhizoids
- reproduce by forming gametes on leafy stems and spores in capsules

39. _____
- simple vascular plants
- reproduce by producing spores in cones in sporangia
- have stems and roots

40. _____
- vascular plants with large leaves called fronds
- grow in moist places; need low light
- reproduce by forming new plants from tips of the fronds

Plants

41. Which plants are vascular plants?
(Circle one or more answers.)

green beans   cypress trees
horsetails   onions
ferns   peat moss
pumpkin plants liverworts

42. Which plants are seed plants?
(Circle one or more answers.)

oak trees   fir trees
ferns   club moss
algae   dandelions
beets   daffodils

43. Which plants are NOT gymnosperms?
(Circle one or more answers.)

lily   pine tree
cypress tree lilac bush
aspen tree   spruce tree

44. Which plants are angiosperms?
(Circle one or more answers.)

cabbage   pine tree
grass   corn plant
roses   palm tree

Write the letter of the correct flower part to answer each question.

_____ 45. Which part of the flower produces the pollen?

_____ 46. Where does fertilization take place?

_____ 47. Which part is sticky for the purpose of holding the pollen?

_____ 48. Which part attracts pollen-carrying insects to the flower?

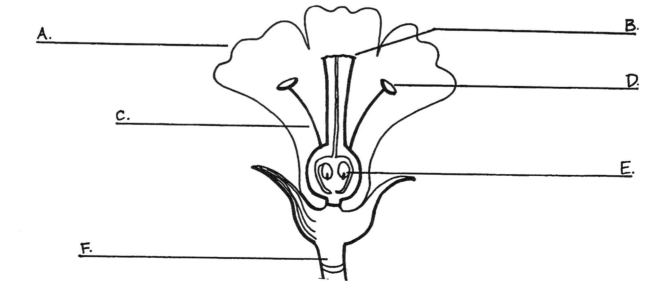

Name _____

Plants

Professor Moses is lecturing about concepts of plant life.
The student has a good list of the concepts, but no understanding of them yet.
Match the professor's explanations with the student's list.
Write a letter from one of the explanations next to the correct concept.

A. The male gametophyte (sperm) is contained in a pollen grain. It is transferred to the female gametophyte (egg).

B. A sperm and an egg join, resulting in a zygote in the ovule.

C. This young plant is growing within the seed. It is the sporophyte stage in the life cycle of a seed plant.

D. In this method of reproduction in simple plants, materials from two different cells come together to form a zygospore.

E. Many plant embryos have a resting stage before they complete their growth and development.

F. This is the name given to the food stored in the seeds of monocots. You may not know this, but it is also the source of many foods that you eat.

G. Water, oxygen, and the right temperature must be present for a seed to do this successfully. Without those, the embryo in the seed will not be able to grow into a young seedling.

H. In this form of asexual reproduction, a new plant is grown from part of another plant. Grafting, making plant cuttings, or separating bulbs from an original plant are all ways to do this.

_____ 49. embryo
_____ 50. fertilization
_____ 51. vegetative propagation
_____ 52. conjugation
_____ 53. endosperm
_____ 54. pollination
_____ 55. germination
_____ 56. dormancy

57. What functions does humus serve in soil?
(*Circle one or more answers.*)
   a. The decayed organic material provides nutrients for plants.
   b. Humus increases the capacity of the soil to hold water.
   c. The plant needs humus in order to carry out photosynthesis.
   d. The humus breaks up the rocks in the soil.

58. Why is loam an ideal soil for farming?
   a. Loam is a natural fertilizer.
   b. Loam is mostly sand, which allows water to flow through the soil easily.
   c. Loam is made of different size particles and allows for air and water to mix in easily.
   d. Loam has a high content of the clay that plants need.

59. Look at the diagram. Label each arrow *carbon dioxide* or *oxygen*.

The Carbon Dioxide-Oxygen Cycle

60. Which is not part of the nitrogen cycle?
   a. Nitrogen-fixing bacteria in the soil remove nitrogen from the air and form nitrates.
   b. Plants absorb nitrates from the soil.
   c. Plants make nitrogen compounds with the nitrates they absorb.
   d. Animals make nitrates and breathe them back into the air.
   e. Animals take in nitrogen compounds when they eat plants.
   f. Bacteria in the soil break down nitrates and return nitrogen to the air.

Life Sciences, Test #4

ANIMALS

Name _____

Date _____

Possible Correct Answers: 90

Your Correct Answers: _____

Students of zoology are studying the system for classifying animals. They have identified the nine main phyla of animals by making a list of characteristics for each. Match the name of the phyla with the description of the organisms in it.

1. _____
- ✓ soft, segmented body
- ✓ bilateral symmetry
- ✓ simplest organisms with a circulatory system

2. _____
- ✓ live in water
- ✓ thick sac of cells with pores, chambers, & canals
- ✓ irregular shapes; no symmetry

3. _____
- ✓ jointed legs
- ✓ segmented bodies
- ✓ hard exoskeleton on outside
- ✓ bilateral symmetry
- ✓ specialized mouthparts for biting, sucking, or chewing

4. _____
- ✓ hollow body surrounded by cells
- ✓ radial symmetry
- ✓ tube or sac-like mouth surrounded by tentacles
- ✓ stinging cells on the ends of tentacles
- ✓ found in water, mostly oceans

5. _____
- ✓ soft bodies usually covered by shell
- ✓ strong, muscular "foot" underneath
- ✓ circulatory system with heart
- ✓ bilateral symmetry

Platyhelminthes
(flatworms)

Chordates

Poriferates

Annelids

Arthropods

Coelenterates

Mollusks

Nematodes
(roundworms)

Echinoderms

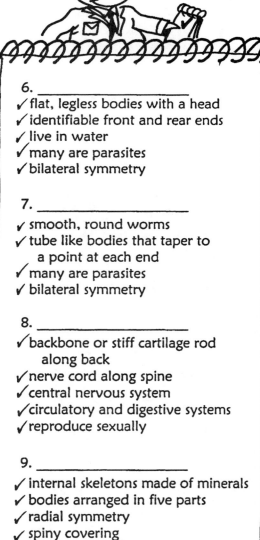

6. _____
- ✓ flat, legless bodies with a head
- ✓ identifiable front and rear ends
- ✓ live in water
- ✓ many are parasites
- ✓ bilateral symmetry

7. _____
- ✓ smooth, round worms
- ✓ tube like bodies that taper to a point at each end
- ✓ many are parasites
- ✓ bilateral symmetry

8. _____
- ✓ backbone or stiff cartilage rod along back
- ✓ nerve cord along spine
- ✓ central nervous system
- ✓ circulatory and digestive systems
- ✓ reproduce sexually

9. _____
- ✓ internal skeletons made of minerals
- ✓ bodies arranged in five parts
- ✓ radial symmetry
- ✓ spiny covering
- ✓ tube feet like suction cups

10–28. Write the number of each animal on the chart to show its phylum.

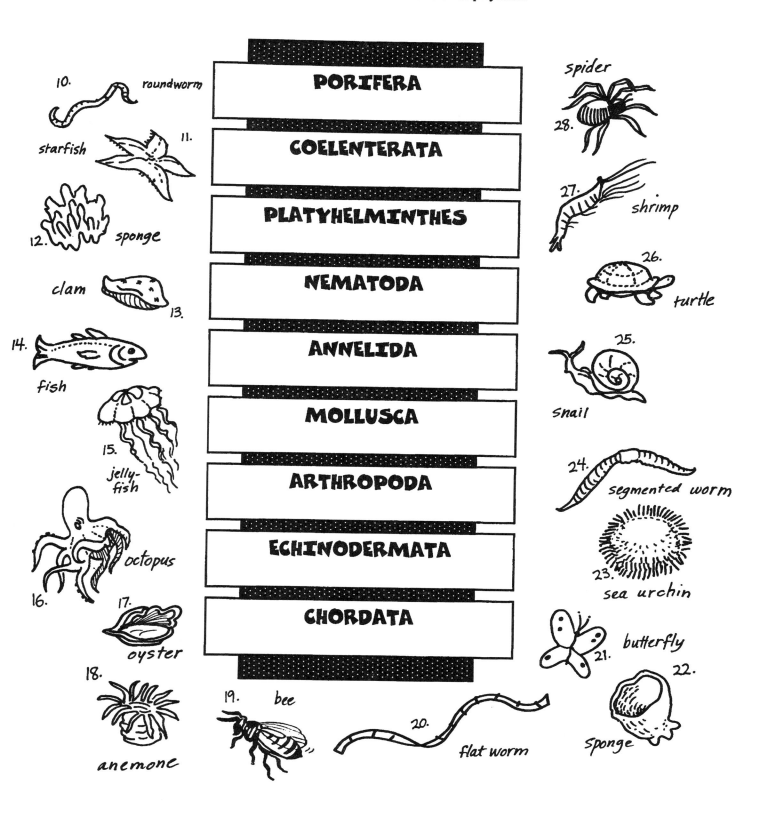

Animals

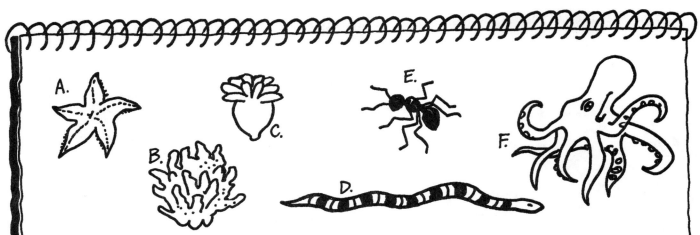

29. Which animals have bilateral symmetry?

_____
_____

30. Which animals have radial symmetry?

_____
_____

31. What are the two main groups of animals (based on the existence of a backbone)?

_____   _____

32. Which is true?
    a. About 50% of all animals are mammals.
    b. About 95% of all animals are invertebrates.
    c. There are more chordates than arthropods.
    d. About 95% of all animals are vertebrates.

33. Which class of animals is most numerous in number of species?
    a. leeches   b. fish   c. insects   d. reptiles

34. The simplest animals are
    a. sponges.   c. roundworms.   e. anemones
    b. insects.   d. mollusks   f. flatworms

35. The process during which insects shed their exoskeletons as they grow is
    a. regeneration.   c. molting
    b. metamorphosis. d. reproduction.

There are many classes of Arthropods. Examples of five classes are used for items 36-45. Match each characteristic or each animal with the correct class. Write the code letters given to the class. Each class will be used more than once.

AR = Arachnids
IN = Insects
CH = Chilopods
DI = Diplopods
CR = Crustaceans

____ 36. three body regions, three pairs of legs, one pair of antennae
____ 37. flat, segmented body, one pair of legs per segment
____ 38. two body regions, four pairs of legs, no antennae
____ 39. gills, two pairs of antennae, hard, flexible exoskeleton
____ 40. round, segmented body, two pairs of legs per segment

____ 41. grasshopper
____ 42. scorpion
____ 43. centipede
____ 44. crab
____ 45. millipede

Use the labels shown to label the stages in the metamorphosis of the butterfly and the grasshopper. A label may be used more than once.

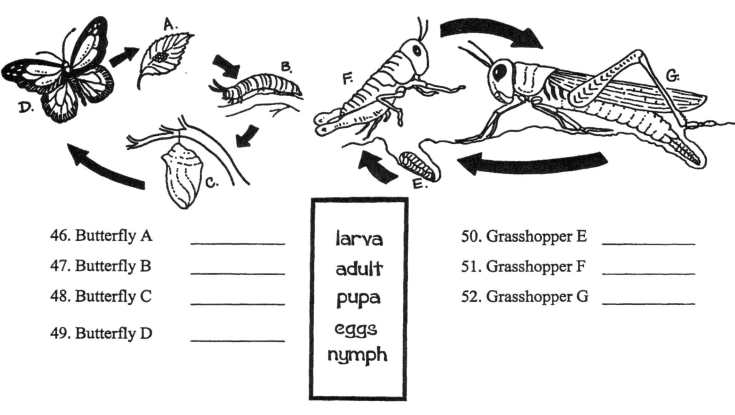

46. Butterfly A _____
47. Butterfly B _____
48. Butterfly C _____
49. Butterfly D _____

larva
adult
pupa
eggs
nymph

50. Grasshopper E _____
51. Grasshopper F _____
52. Grasshopper G _____

Animals

Phylum Chordata has seven classes of vertebrates.

For each animal characteristic described below, write the name of all the chordate classes that have the characteristic in most animals of the class. (Combine the three different fish classes into one class labeled 'fish'.)

Which class (or classes) . . .

53. . . . are cold-blooded? _____
54. . . . live mostly in water? _____
55. . . . breathe air through lungs? _____
56. . . . reproduce with external fertilization? _____
57. . . . have moist skin with no scales? _____
58. . . . produce milk for their young? _____
59. . . . have hair or fur body-covering? _____
60. . . . breathe with gills? _____
61. . . . have a 3-chambered heart? _____
62. . . . have hollow bones? _____
63. . . . have moist skin with scales? _____
64. . . . have a 4-chambered heart? _____
65. . . . live mostly on land? _____
66. . . . are warm-blooded? _____
67. . . . change appearance as they develop? _____
68. . . . have wings, feathers, and beaks? _____
69. . . . have a 2-chambered heart? _____
70. . . . have dry skin with scales? _____
71. . . . reproduce with internal fertilization? _____
72. . . . can live on land or in water? _____

AMPHIBIANS

REPTILES

FISH

MAMMALS

BIRDS

73. Which statements are true about the reproductive process of budding? (Circle one or more.)
    a. A tiny piece of tissue breaks off from the organism and develops into a new organism.
    b. Budding is a form of reproduction common to sponges.
    c. In budding, sex cells are contributed by two parents.

74. Which is an example of regeneration?
    a. A piece of tissue breaks off from a hydra. The piece grows into an adult hydra.
    b. A starfish loses an arm. After a while, a new arm grows in place of the old one.
    c. An earthworm produces both an egg and a sperm. The sperm fertilizes the egg.

75. Which statement is NOT true about external fertilization?
    a. External fertilization works well among animals that live in water.
    b. Most fish, birds, and reptiles reproduce by external fertilization.
    c. External fertilization means that the sperm fertilize eggs outside the female's body.

76. When internal fertilization occurs, there is a period of time between the fertilization of the egg and the birth of the offspring. This period is called
    a. estrus   b. lactation   c. gestation   d. contraction

A dog barks fiercely and growls when a stranger enters the yard. A group of birds flap their wings wildly when a strange bird comes into their space.

77. These are examples of the animal behavior of
    a. communication.
    b. territoriality.
    c. adaptation.
    d. hibernation.

78. Which is NOT an example of an animal response to a stimulus?
    a. A cat sees the movement of a mouse. The cat pounces.
    b. A child touches a thorn on a rose. She jerks her hand away.
    c. A hungry dog sees a dish full of food.

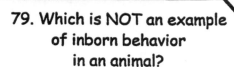

79. Which is NOT an example of inborn behavior in an animal?

a. Your mouth waters at the site of a chocolate bar.
b. A spider spins a web for its home.
c. A baby turtle breaks out of its shell.
d. A rat learns to find its way through a maze to a dish of food.

80. Which is NOT an example of acquired behavior in an animal?

a. The pupils of a deer's eyes become small when headlights shine in them.
b. A baby finds a rattle her mother hid.
c. A dog gets his owner when the phone rings.
d. A baby bear learns how to find food from his mother.

81. Which is NOT an example of social behavior in an animal?

a. Birds travel in flocks when they migrate.
b. A rat eats poison and gets ill. He stays away from that food from then on.
c. Wolves live together in packs.
d. A male duck does a kind of courtship dance on the water.

82–83: Circle one or more answers for each item.

82. Fiddler crabs change color at daybreak and at sunset.
Bears hibernate during the winter.
People get jet lag when they fly across several time zones.
These things happen because of
 a. a living clock in the animals.
 b. acquired behavior.
 c. conditioning.

83. Territoriality helps animals survive by
 a. reducing the competition for food.
 b. reducing the rates of reproduction.
 c. changing the kinds of predators.
 d. reducing the competition for mates.
 e. increasing the population of the predators.

Write T (true) or F (false) for each statement.

HEREDITY

84. RNA is the genetic material in the cell that controls every inherited trait.

85. A gene is the basic unit of inheritance passed from parents to offspring.

86. Gregor Mendel discovered basic laws of heredity by studying corn plants.

87. A recessive form of a trait prevents the dominant form from showing.

88. An offspring gets half its genes from each parent.

The Punnett square below shows the possible ways that genes can combine for the trait of tallness (height) in a plant.

The **T** stands for the dominant gene for tallness. The **t** stands for the recessive gene (the gene for Not tall-or shortness). Pure tall is labeled **TT**. Pure short is labeled **tt**. **Tt** is hybrid tall.

One parent is **TT** (pure tall), with both genes dominant.

The other parent is **Tt** (hybrid tall) with one dominant and one recessive gene.

89. Finish the Punnett square by filling in the results.

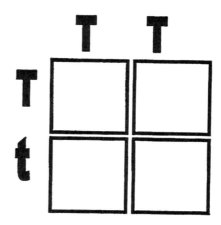

90. Which of these describes the offspring?
   a. 25% pure tall, 50% hybrid tall, 25% pure short
   b. 50% pure tall, 50% pure short
   c. 100% hybrid tall
   d. 100% pure tall
   e. 50% pure tall, 50% hybrid tall

Life Sciences, Test #5

ECOLOGY

Name _____

Date _____

Possible Correct Answers: 35

Your Correct Answers: _____

1. A population of beetles has a place where the beetles live within the forest ecosystem. This place is their _____.

2. The part of the biosphere that surrounds a seahorse is the animal's _____.

3. All the organisms living together in a forest pond are the pond's _____.

4. All the prairie dogs that are living in a grassland community are known as the prairie dog _____.

5. In a desert, the sagebrush species has a special role, or _____ in the environment.

6. There is a maximum number of moles that can be supported by a particular habitat. This number is the habitat's _____.

ecology   community
environment   niche   compete   carrying capacity
habitat   succession   ecosystem
        population   climax community

Ecology

7. Lichen is made up of fungi and algae. Both organisms benefit each other. The fungi provide shelter and a moist environment for the algae. The algae make food for the fungi.

This relationship is
   a. parasitism.   c. commensalism.
   b. mutualism.   d. competition.

8. A tapeworm lives inside an animal. The tapeworm gets water, oxygen, and food from the animal, but is harmful to its host. The animal gets no benefit from the relationship.

This relationship is
   a. parasitism.
   b. mutualism.
   c. commensalism.
   d. competition.

9. The remora fish attaches itself to a shark. The remora travels around with the shark, eating leftover scraps of food. The remora does not help or harm the shark.

This relationship is
   a. parasitism.   c. commensalism.
   b. mutualism.   d. competition.

10. A fungus on a log helps the log decay. The fungus is a
   a. scavenger.   c. parasite.
   b. decomposer.   d. producer.

11. A vulture feeds on a dead skunk. The vulture is a
   a. scavenger.   c. decomposer.
   b. parasite.   d. producer.

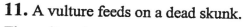
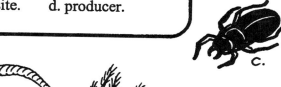

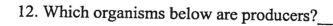

12. Which organisms below are producers?_____

13. Which organisms below are consumers?_____

Ecology

14. Which organisms in the food chain are producers?
_____

15. Which are consumers?
_____

16. Which are primary consumers?
_____

17. Which are secondary consumers?
_____

plants and grasses ← grasshopper ← frog ← garter snake

18-20: Write the name of a biome to match the features.

fresh water
salt water
tropical rain forest
grassland
desert
temperate deciduous forest
taiga
tundra

18. In which biome would these features be found?

*foxes   squirrels   bears
oak trees   woodpeckers*

Answer _____

19. In which biome would these features be found?

*crayfish   trout   marshes
snakes   alligators*

Answer _____

20. In which biome would these features be found?

*irregular precipitation   grasses
antelopes   gazelles   prairie dogs*

Answer _____

21. Which biome is pictured?

_____

22. The size of a population in the environment could be limited by
   a. the food supply.
   b. the oxygen supply.
   c. the temperature.
   d. diseases.
   e. the number of predators.
   f. all of the above.
   g. a and e.

23. In a community, the plant and animal populations are gradually replaced by new plant and animal populations. This process is
   a. succession.
   b. commensalism.
   c. symbiosis.
   d. competition.

24. Which are fossil fuels?
(Circle one or more answers.)
   a. nuclear energy   d. petroleum
   b. steam   e. electricity
   c. coal   f. natural gas

25. A biodegradable substance
   a. is a hazardous chemical waste.
   b. destroys the environment.
   c. can be decomposed into substances that do not harm the environment.

26. Which of these are nonrenewable resources? *(Circle one or more answers.)*

timber **grass** **natural gas** **coal**

crops **topsoil** **petroleum** **animals**

Ecology

Each circle describes a natural cycle.
Each description is missing one phase or part.
Write a phrase or sentence that fills in the missing part. (Write in the box.)

27. Carbon Dioxide-Oxygen Cycle

Plants take in carbon dioxide during photosynthesis.
Plants use the carbon to make food.
During photosynthesis, plants release oxygen into the air.

[ ]

Animals give off carbon dioxide during respiration.
Dead plants and animals give off carbon dioxide as they decay.

28. Water Cycle

Water from the soil, oceans, rivers, and lakes evaporates and turns into water vapor in the air.
Water vapor rises and cools, condensing to tiny drops of water or ice in clouds.

[ ]

Most water returns to the ocean or settles into the ground.
Plants drink water and release it into the air through transpiration.

29. Nitrogen Cycle

Plants and animals need nitrogen, but cannot take them from the air.

[ ]

Plants take the nitrates in through their roots.
Other bacteria in the soil take the nitrates and release nitrogen back into the air.

Ecology

30. Building the Alaska Pipeline

A. Describe one benefit. _____
_____

B. Describe one consequence to the environment.
_____
_____
_____

31. Clear cutting in forests

A. Describe one benefit. _____
_____

B. Describe one consequence to the environment.
_____
_____
_____

32. Describe a harmful effect of thermal pollution. _____
_____

33. Define acid rain and describe its effect on the environment.
_____
_____

Choose two of these techniques. For each one, describe how its use is helpful to the environment.

crop rotation
contour plowing & planting
reforestation
recycling
sewage treatment
sound-absorbent building materials
burying utility wires and lines
open-space planning

34. _____
_____
_____

35. _____
_____
_____

Human Body & Health Skills Checklists

Human Body & Health Tests # 1:
BODY SYSTEMS, PART I
SKELETAL SYSTEM-MUSCULAR SYSTEM, NERVOUS SYSTEM, ENDOCRINE SYSTEM
Test: pages 69–75

Skill	Test Items
Recognize the functions of different body systems	1–9
Distinguish between tissue, organ, and system	10
Identify different kinds of cells	13–16
Identify components of body systems	17–23
Identify major bones of the human skeleton	24–37
Show understanding of the components and working of the skeletal-muscular system	38–46
Recognize different kinds of joints in the body	47–51
Identify components of the nervous system and their functions	52–59
Label and describe a neuron; describe how impulses travel from one neuron to another	58–59
Recognize the function and location of different taste buds	60–62
Identify structures of the ear and their functions	63–68
Show understanding of the sense of touch and its relationship to nerve endings in the skin	69–70
Identify structures of the eye and their functions	71–75
Recognize organs and functions of the endocrine system	76–80

Human Body & Health Tests # 2 :
BODY SYSTEMS, PART II
CIRCULATORY SYSTEM, RESPIRATORY SYSTEM, DIGESTIVE SYSTEM. REPRODUCTIVE SYSTEM

Test Location: pages 76–81

Skill *Test Items*

Recognize structures of the circulatory system and their functions 1–7, 10, 11
Show understanding of the workings of the circulatory system 1–7, 10, 11
Recognize components of the blood and their different functions 8, 9
Understand the concept of blood type, donation, and receiving of blood 12
Differentiate between different types of circulation 13, 14
Show understanding of the working of the heart 15–20
Identify the parts of the respiratory system 21–28
Recognize the components of the respiratory system and their functions 29–32
Understand how the respiratory system works 33
Show understanding of the organs
    and function of the excretory system 34–36, 39–40
Identify parts and function of the integumentary system 37–38
Recognize the components of the digestive system and their functions 41–49
Recognize components of the reproductive system and their functions 50–59
Show understanding of the occurrence of twins and of sex determination 59–60

Human Body & Health Test # 3
DISEASES & DISORDERS
Location: pages 82–85

Skill *Test Items*

Show understanding of communicable diseases.
 Identify some communicable diseases and their causes 1, 3, 5
Show understanding of inherited diseases 2
Recognize the function of body systems to prevent and fight disease 4, 22
Show understanding of the concepts of carrier, immunity, and epidemic 6–7
Recognize different kinds of diseases and their symptoms 9–21
Distinguish between diseases caused by viruses and those caused by bacteria 10, 12
Recognize diseases that fall into categories of:
 mental illness, congenital disease,
 degenerative disease, and environmental disease 11, 13, 15, 17
Show understanding of the development and uses of vaccines 22
Show understanding of how the body works naturally
 to fight disease and infections 23
Identify the role of the FDA 24
Show understanding of some treatments used for common diseases 25, 26
Show understanding of ways to stop or slow the spread of disease 27
Show understanding of uses, dangers, and body responses to drugs 28–35

Human Body & Health Test # 4
FITNESS & HEALTH
Test Location: pages 86–91

Skill *Test Items*

Explain the benefits of good health and fitness behaviors 1–12
Show understanding of the nutrients needed and used by the body,
 their sources, and their functions in the body 13–18
Identify benefits and sources of nutrients needed by the body 19–22, 25–29
Recognize ways the body uses food 14–16, 18, 23, 24
Identify different kinds of exercise and the benefit of each to the body 30–33
Recognize the benefits of exercise 34, 36
Distinguish between low-density and high-density lipoproteins 37
Recognize body responses to stress and methods for stress management 38, 39
Describe appropriate responses to emergencies 40–47
Show familiarity with the effects of various drug substances on the body
 and the dangers of use or abuse of those substances. 48–70

Human Body & Health, Test #1

SKELETAL SYSTEM—MUSCULAR SYSTEM—
NERVOUS SYSTEM—ENDOCRINE SYSTEM

Name _____ Possible Correct Answers: 80

Date _____ Your Correct Answers: _____

Professor Felicity Femur is teaching her students about systems of the human body. She is describing the functions and purposes of the different systems. Write the number of the matching description on the line in front of the system name.

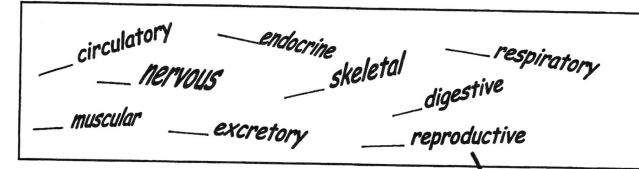

1. gets rid of waste products in the body

2. enables humans to produce offspring

3. transport system that carries blood, nutrients, and other materials through a system of vessels

4. a series of ductless glands that produce hormones, chemicals to control many body functions

5. changes food to a form that can be used by the body cells

6. controls muscles and regulates body activities through a system of messages carried between the brain and the rest of the body

7. gives shape and strength to the body enables all body movements

8. supplies oxygen to the cells and removes carbon dioxide from the blood

9. internal frame that supports and shapes the body and protects internal organs

Middle Grade Book of Science Tests   Copyright ©2001 by Incentive Publications, Inc., Nashville, TN.

Body Systems, Part I

10. _____ = group of cells that have a similar shape and function

11. _____ = group of tissues that work together to carry out a specific function

12. _____ = group of organs working together to carry out one or more functions

Write the correct label for each group of cells: **muscle, bone, blood, nerve.**

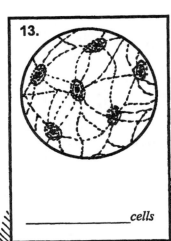
13.
_____ cells

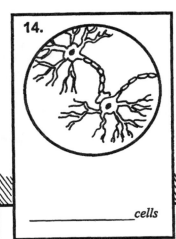
14.
_____ cells

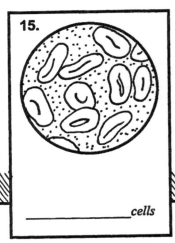
15.
_____ cells

16.
_____ cells

17. Which is NOT a part of the respiratory system?
   a. trachea
   b. dendrites
   c. alveoli
   d. bronchi

18. Which is NOT a part of the circulatory system?
   a. stomach
   b. aorta
   c. capillaries
   d. platelets

19. Which is NOT a part of the digestive system?
   a. spleen
   b. cerebellum
   e. teeth
   d. esophagus

20. Which is NOT a part of the nervous system?
   a. axons
   b. pancreas
   c. neurons
   d. medulla

21. Which is NOT a part of the excretory system?
   a. skin
   b. lungs
   c. mandible
   d. kidneys

22. Which is NOT a part of the endocrine system?
   a. thyroid
   b. pituitary gland
   c. insulin
   d. hemoglobin

23. Which is NOT a part of the muscular system?
   a. marrow
   b. tendons
   c. deltoid
   d. ligaments

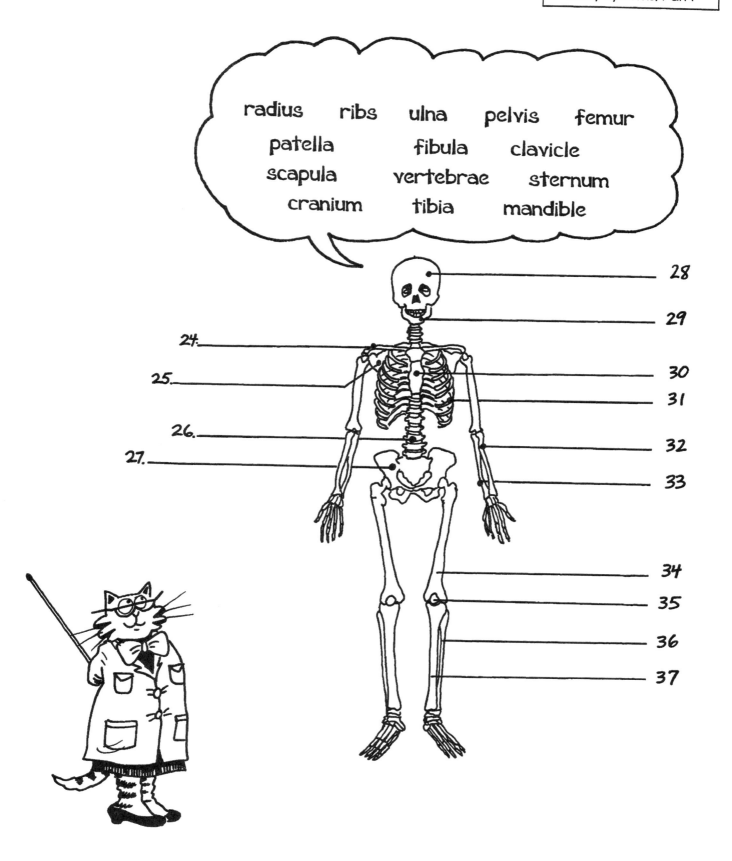

Body Systems, Part I

38. Which statements about bones are NOT true?
(Circle one or more answers.)

a. Some bones are hard, and some bones are spongy.

b. Bones are covered with a thick membrane called the diaphragm.

c. Bones are made of nonliving minerals.

d. Blood vessels carry food and oxygen to bone cells.

e. Blood cells are produced by red bone marrow.

f. Red marrow is only found in ribs.

Circle the correct word to complete the sentence.

39. Bones are held together by (tendons, ligaments).

40. Most joints are padded with (cartilage, muscles).

41. Skeletal muscles are (smooth, striated) muscles.

42. (Ligaments, Tendons) attach muscle fibers to the bones.

43. A muscle works by (contracting, relaxing).

44. (Smooth, Striated) muscles are voluntary muscles.

45. The heart muscle is (a voluntary, an involuntary) muscle.

46. The walls of blood vessels and of the digestive tract contain (smooth, striated) muscle.

Use the words at the right to label each joint and give an example.

cranium, elbow, hip, spine, wrist

pivot, ball & socket, hinge, fixed, sliding joint

47. _____

48. _____

49. _____

50. _____

51. _____

Which part of the nervous system.....

52. .... controls memory, senses, and speech?

  A. cerebrum   B. cerebellum   C. medulla

53. .... controls balance, coordination, and muscle activity?

  A. cerebrum   B. cerebellum   C. medulla

54. .... carries impulses from receptors to the central nervous system?

  A. motor neurons   B. interneurons   C. sensory neurons

55. .... controls involuntary actions such as breathing and swallowing?

  A. cerebrum   B. cerebellum   C. medulla

56. .... carries impulses away from the central nervous system to the muscles?

  A. motor neurons   B. interneurons   C. sensory neurons

57. carries messages between the brain and spinal cord and the rest of the body?

  A. the central nervous system   B. the peripheral nervous system

Use the terms nucleus, axon, dendrite, and synapse for 58 and 59.

58.

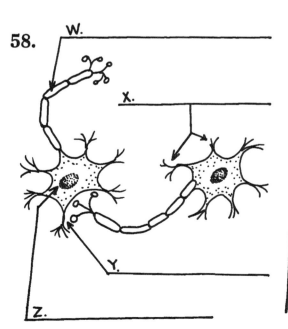

W.
X.
Y.
Z.

59. Describe how an impulse travels from one neuron to the other.

Body Systems, Part I

60. Max takes a bite of fudge. Which part of his tongue will the taste bud sense the sweetness?
 a. the front
 b. the sides
 c. the middle
 d. the back

61. Alphonso bites into a lemon. Which part of his tongue will sense the sour taste?
 a. the front
 b. the sides
 c. the middle
 d. the back

62. The receptors for smell are located in the
 a. sinus glands.
 b. olfactory membrane.
 c. epiglottis.
 d. tongue.

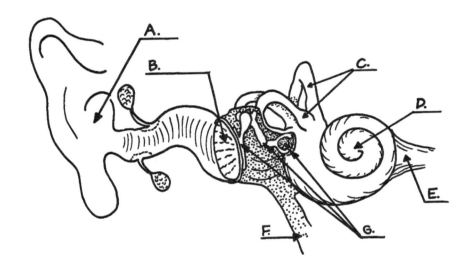

Read each description of the structure and function of a part of the ear.
Write the letter from the diagram which points to that part.

___ 63. a tightly-stretched membrane that separates the outer ear from the inner ear; vibrates when sound hits it

___ 64. passageway from the back of the nose that allows air to pass into the middle ear

___ 65. funnel-shaped structure that directs the sound into the ear

___ 66. a fluid-filled, coiled tube that produces nerve impulses in response to sound waves

___ 67. tiny bones which pass the vibrations on to the inner ear

___ 68. carries the impulses from the receptor cells in the cochlea to the brain

69. Nerve endings in the skin that sense heat, cold, pressure, and pain are located in
   a. the epidermis.
   b. the fat cells.
   c. the dermis.
   d. the oil glands.

70. The soles of the feet are very sensitive body parts because
   a. they receive much pressure from the person's weight.
   b. they are rarely in the sun.
   c. they have a rich supply of nerve endings.
   d. the skin on them is unusually thin.

71. The light-sensitive structure in the eye that catches images is the _____.

72. The size of the pupil is controlled by the muscular _____.

73. The _____ carries the images from the eye as signals to the brain.

74. The tough protective covering of the eye is the _____.

75. The black hole where light enters the eye is the _____.

76-80. Five of these statements about the endocrine system are not quite accurate. Circle the letters and make them correct by changing a word or phrase.

   A. The pancreas produces adrenalin, which controls the amount of sugar in the blood.
   B. Meningitis is a disease that results from too little insulin released in the body.
   C. The pituitary gland controls the release of hormones from many other glands.
   D. When adrenaline is released into the body, muscle activity decreases.
   E. The hormone insulin controls the storage of glycogen in the liver.
   F. Hormones are chemicals that control many body functions.
   G. The glands in the endocrine system produce hormones.
   H. Each hormone affects many different kinds of tissues.
   I. The master gland in the system is the thyroid.
   J. The adrenal glands regulate body metabolism.

Human Body & Health, Test #2

CIRCULATORY SYSTEM—RESPIRATORY SYSTEM— DIGESTIVE SYSTEM REPRODUCTIVE SYSTEM

Name _____

Date _____

Possible Correct Answers: 60

Your Correct Answers: _____

1. What vessels carry blood away from the heart? _____
2. What vessels carry blood to the heart? _____
3. What structures link arteries and veins? _____
4. What chemical in red blood cells carries oxygen to tissues? _____
5. Why are veins blue? _____
6. Why do veins have valves? _____
7. Why are arteries thicker than veins? _____

8. Which part of the blood will surround an infected area in the body?
   a. plasma   b. red blood cells   c. white blood cells   d. platelets

9. Which part of the blood helps the blood clot?
   a. plasma   b. red blood cells   c. white blood cells   d. platelets

10. What is NOT true?
a. The capillaries exchange food and oxygen for wastes and carbon dioxide from the tissues.
b. Capillaries help to control body temperature.
c. Capillaries are not found in the digestive organs.

11. Which is NOT true?
a. Each time the heart contracts, the arteries expand, and each time the heart relaxes, the arteries contract.
b. Blood moves more slowly through arteries than through veins.
c. A pulse is caused by the expansion and contraction of the arteries.

12. After an accident, a person needs a blood transfusion. She cannot receive type A blood. What is her blood type?
   a. Type AB b. Type B or O c. Type A d. Type A or AB.

13. The movement of blood around the tissues of the heart is
   a. pulmonary circulation.
   b. coronary circulation.
   c. systemic circulation.

14. The flow of blood through the heart, to the lungs, and back to the heart is
   a. pulmonary circulation.
   b. coronary circulation.
   c. systemic circulation.

Examine the diagram of the heart. Write the letter of the matching arrow to describe the path of the blood.

_____ 15. The superior vena cava brings low-oxygen blood from the body into the right atrium.

_____ 16. The right atrium contracts and forces blood through the valve into the right ventricle.

_____ 17. The right ventricle contracts, forcing blood through the pulmonary arteries toward the lungs.

_____ 18. The oxygen-rich blood flows back from the lungs through the pulmonary veins into the left atrium.

_____ 19. The left atrium contracts, forcing the blood into the left ventricle.

_____ 20. The heart pumps the blood out of the left ventricle through the aorta to the body.

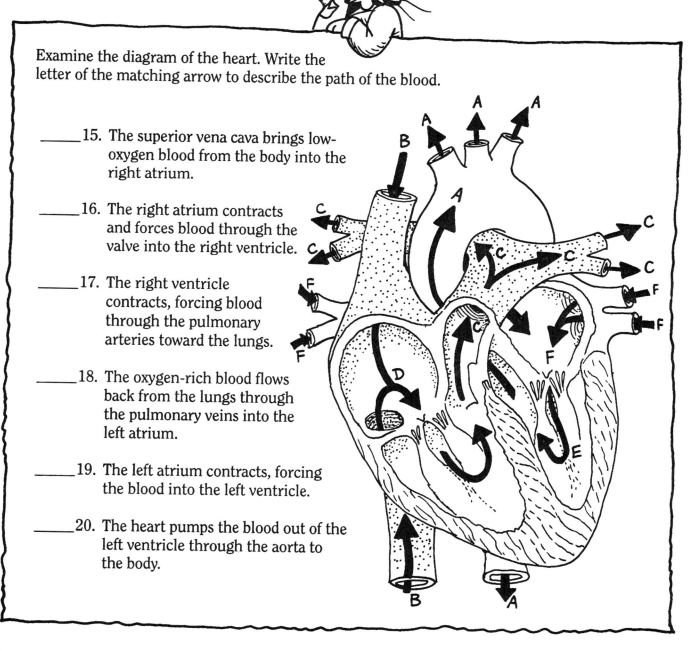

Body Systems, Part II

21–28: Label the parts of the respiratory system indicated on the diagram.

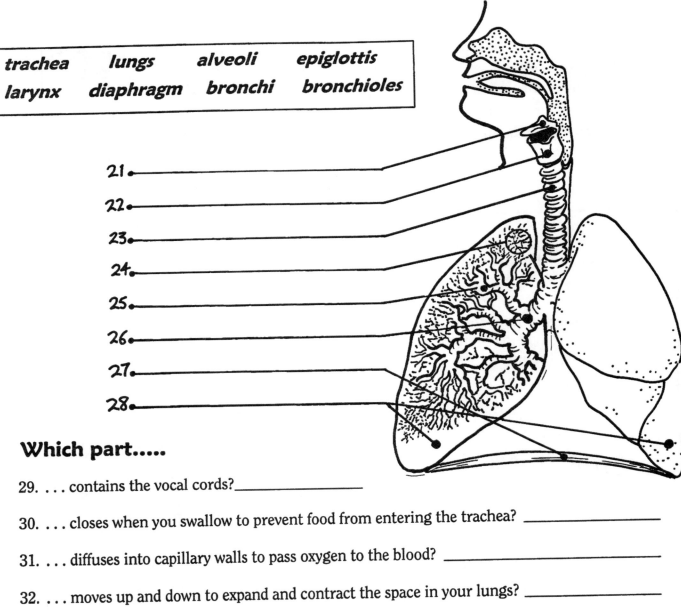

trachea   lungs   alveoli   epiglottis
larynx   diaphragm   bronchi   bronchioles

21. _____
22. _____
23. _____
24. _____
25. _____
26. _____
27. _____
28. _____

Which part.....

29. ... contains the vocal cords? _____

30. ... closes when you swallow to prevent food from entering the trachea? _____

31. ... diffuses into capillary walls to pass oxygen to the blood? _____

32. ... moves up and down to expand and contract the space in your lungs? _____

33. ... Which of these happen when you inhale?
(Circle one or more answers.)
   a. Many alveoli are filled with air.
   b. The diaphragm flattens and moves down.
   c. The volume of the chest cavity decreases.
   d. The diaphragm relaxes.
   e. Air pressure in the lungs increases.
   f. The volume of the chest cavity increases.
   g. Carbon dioxide is expelled through the trachea.

34. Which of these does NOT remove wastes?
   a. kidneys
   b. liver
   c. gall bladder
   d. lungs
   e. skin

35. What is formed when the kidneys remove waste products from the blood?
   a. bile
   b. urine
   c. insulin
   d. water vapor

36. What is formed when the liver removes chemical waste from the blood?
   a. bile
   b. urine
   c. insulin
   d. blood

37. What is the function of the integumentary system?
   a. to help process liquids.
   b. to neutralize poisons in the body.
   c. to cover and protect the body.
   d. to strengthen the immune system.

38. Which is NOT part of the integumentary system?
   a. white blood cells
   b. skin
   c. hair
   d. fingernails
   e. toenails

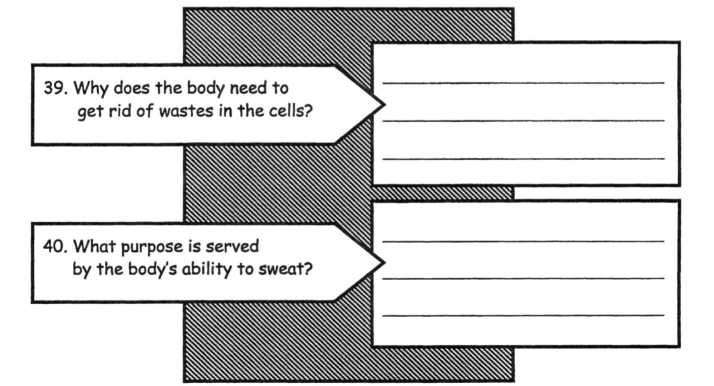

39. Why does the body need to get rid of wastes in the cells?

40. What purpose is served by the body's ability to sweat?

Body Systems, Part II

Professor Femur is explaining how her favorite lunch, a nice big fat tuna fish sandwich, makes its way through the digestive system.

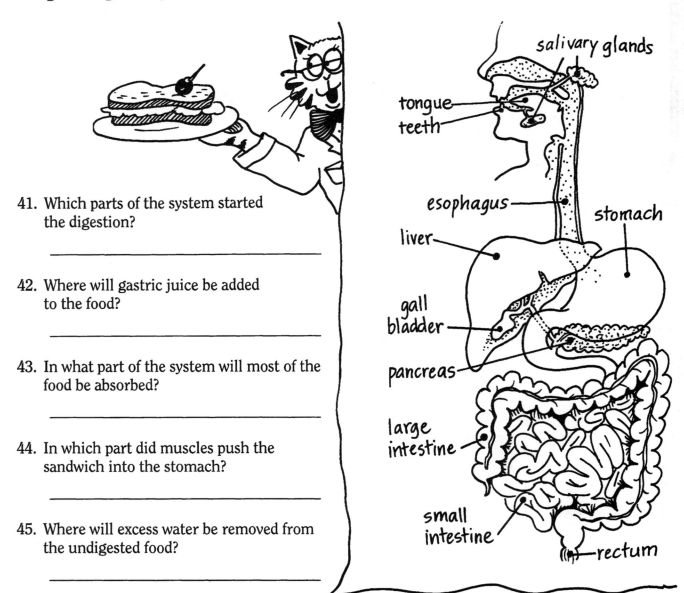

41. Which parts of the system started the digestion? _____

42. Where will gastric juice be added to the food? _____

43. In what part of the system will most of the food be absorbed? _____

44. In which part did muscles push the sandwich into the stomach? _____

45. Where will excess water be removed from the undigested food? _____

46. From what part will bile come to add to the digestion of the food? _____

47. What part will help remove undigested food out of the body? _____

48. How will the pancreas help with digestion of the sandwich? _____

49. When bile is added to the food, how will the bile help digestion? _____

Name _____

Body Systems, Part II

Write the number of the description to match each reproductive structure or process.

50. the release of an egg from ovary

51. the male sperm-producing organs

52. the moment of fertilization

53. the female egg-producing organs

54. the time a human is first able to reproduce

55. a fertilized egg which has passed into the uterus and begun to divide

56. the hollow, thick-walled organ that contains, supports, and nourishes a growing fetus

57. a tissue that forms within the wall of the uterus, supplying blood and food to the growing fetus

58. the structure that transports oxygen and wastes to and from the embryo

___ embryo   ___ PUBERTY

___ ovulation

___ ovaries

___ umbilical cord

___ uterus

___ conception

___ testes

___ placenta

59. Explain the difference between the occurrence of identical twins and fraternal twins.

_____

_____

_____

60. An egg is fertilized by a sperm that carries one X-chromosome.
    a. The offspring will be a boy.
    b. The offspring will be a girl.
    c. The offspring will be fraternal twins.
    d. The offspring could be either a boy or a girl.

Human Body & Health, Test #3 — page 1 of 4

DISEASES & DISORDERS

Name _____

Date _____

Possible Correct Answers: 35

Your Correct Answers: _____

1. Professor Femur is doing research on communicable diseases. Which of these diseases would NOT be a part of her study? (Circle one or more answers.)

   a. influenza
   b. athlete's foot
   c. diabetes
   d. a food allergy
   e. tuberculosis
   f. AIDS
   g. strep throat
   h. lead poisoning

2. Dr. Femur goes on to study inherited diseases. Which of these diseases might be a part of her study? (Circle one or more answers.)

   a. meningitis
   b. hemophilia
   c. chicken pox
   d. typhus
   e. measles
   f. malaria
   g. mumps
   h. influenza

True or False?

_____ 3. Microbes cause diseases.

_____ 4. Cold weather causes diseases.

_____ 5. Antibodies are made by red blood cells in the body.

Write the correct word to complete the sentence.

6. A _____ can pass on a communicable disease without showing symptoms of the illness.

7. _____ is the ability of a person's body to defend itself against a disease without getting sick from the disease.

8. During an _____, large numbers of people get a disease in a short period of time.

82

Middle Grade Book of Science Tests

Copyright ©2001 by Incentive Publications, Inc., Nashville, TN.

For 9–21, write a letter from one of the professor's reference books.

Where will Dr. Femur look to find out about . . .

_____ 9. . . . causes of strokes?

_____ 10. . . . polio or chicken pox?

_____ 11. . . . a treatment for arthritis?

_____ 12. . . . tuberculosis or pneumonia?

_____ 13. . . . symptoms of schizophrenia?

_____ 14. . . . how to repair a broken wrist?

_____ 15. . . . a disease that a baby has from birth?

_____ 16. . . . an effective chemical substance that prevents the growth of bacteria on living tissue such as skin?

_____ 17. . . . lead poisoning or a lung disease caused by asbestos dust?

_____ 18. . . . substances that can be used to slow or stop the growth of bacteria?

_____ 19. . . . treating a disorder of cell growth, where cells divide without control, damaging healthy cells in the process?

_____ 20. . . . a chemical that can be used to deaden pain or make a patient unconscious during surgery?

_____ 21. . . . a disease where a patient develops antibodies to a normally harmless substance such as feathers or food?

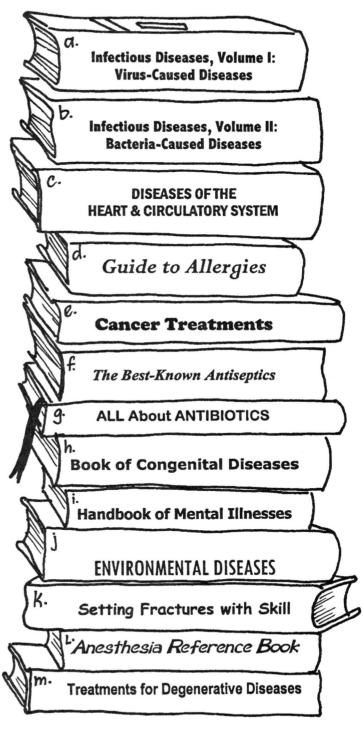

a. Infectious Diseases, Volume I: Virus-Caused Diseases

b. Infectious Diseases, Volume II: Bacteria-Caused Diseases

c. DISEASES OF THE HEART & CIRCULATORY SYSTEM

d. Guide to Allergies

e. Cancer Treatments

f. The Best-Known Antiseptics

g. ALL About ANTIBIOTICS

h. Book of Congenital Diseases

i. Handbook of Mental Illnesses

j. ENVIRONMENTAL DISEASES

k. Setting Fractures with Skill

l. Anesthesia Reference Book

m. Treatments for Degenerative Diseases

Diseases & Disorders

22. Which statement about vaccinations is NOT true?

a. A vaccine causes the body to develop antibodies against a disease.
b. A vaccine causes a natural immunity against the disease.
c. A vaccine is a solution of weakened or dead microbes.

23. Which is NOT one of the body's natural defenses against disease?

a. Mucus in the nose and throat keeps microbes from going further into the body.
b. Acid in the stomach kills microbes that enter the body with food.
c. Unbroken skin keeps disease-causing microbes out of the body.
d. Vaccines in the body create immunity against diseases.
e. Cilia in nasal passages filter microbes to keep them out of the lungs.
f. White blood cells surround microbes and digest them.
g. Lymph cells make antibodies that kill harmful microbes.

24. Which is NOT a role of the FDA?

a. to test the safety of water in public places
b. to approve the safety of food products
c. to investigate the value and effectiveness of new drugs
d. to approve the safety of new drugs

25. Chemotherapy is a technique frequently used to treat

a. AIDS.
b. strep throat.
c. viruses.
d. cancer.

26. Antihistamines are frequently used to treat

a. allergies.
b. sore muscles.
c. arthritis.
d. polio.

27. Choose three of the following substances or practices. Explain how each one can help to prevent the spread of disease.

- disinfectant
- antiseptic
- insect control
- quarantine
- water treatment
- food inspection

A. _____

B. _____

C. _____

Write the number of the description next to the matching term.

___ interaction

___ drug

___ addictive

___ physical dependence

___ withdrawal

___ medicine

___ tolerance

___ psychological dependence

28. any substance that can change life processes in the body

29. drug that is used to treat diseases or symptoms

30. a drug changing the effect of another drug

31. a condition in which the drug user has lost control over the drug use

32. a need to take a drug in order to avoid withdrawal symptoms

33. the body's build-up of resistance to a drug's action

34. a condition in which a drug user has a strong craving and an emotional need to use a drug

35. the sickness that occurs when a person who is physically dependent on a drug stops taking it

Human Body & Health, Test #4

FITNESS & HEALTH

Name _____

Date _____

Possible Correct Answers: 70

Your Correct Answers: _____

Look at the list of practices that contribute to fitness and health.
Write a benefit that a person can receive from each one.

1. Exercise regularly. _____

2. Eat healthy food. _____

3. Keep your body clean. _____

4. Practice good posture. _____

5. Drink plenty of water. _____

6. Learn to manage stress. _____

7. Avoid dangerous situations. _____

8. Value yourself and your body. _____

9. Get plenty of sleep and relaxation. _____

10. Work to make and keep good friends. _____
_____

11. Keep harmful substances out of your body. _____
_____

12. Get regular checkups from your doctor and dentist.
_____

Fitness & Health

13. Eating a balanced diet means
   a. eating an equal amount of several foods each day.
   b. eating the same amount of food daily.
   c. eating correct amounts of nutrients needed for good health.

14. Energy-rich nutrients that are the body's major source of energy are
   a. carbohydrates.
   b. proteins.
   c. vitamins.
   d. minerals.

15. Which does not accurately describe a use for fat in the body?
   a. Stored fat protects your body's internal organs from injury.
   b. The blood needs fat cells in order to flow easily.
   c. Fatty acids are needed to build cell membranes.
   d. Fats provide energy for the body.

16. To get the essential amino acids needed for growth and repair of body tissue, it is important to eat the correct amount of
   a. fats.
   b. vitamins.
   c. calcium.
   d. protein.

17. What does it mean when a food is described as having *"empty calories"*?

_____
_____
_____
_____
_____

18. In order to be used by the body as an energy source, food needs to be broken down into
   a. glucose.
   b. protein.
   c. fat.
   d. carbohydrates.
   e. minerals.
   f. water.

Name

Fitness & Health

Name four major food groups and
a healthy food from each group.

GROUP	FOOD
19. _____	_____
20. _____	_____
21. _____	_____
22. _____	_____

For items 23–29, circle one or more answers.

23. What is the value of eating complex carbohydrates rather than simple carbohydrates?
   a. Complex carbohydrates can provide more fiber in the diet.
   b. Most complex carbohydrates have more nutrients than simple carbohydrates.
   c. Complex carbohydrates can provide vitamins not found in simple carbohydrates.
   d. Most complex carbohydrates have no calories.

24. Most likely, what will happen when someone eats more calories than their body burns?
   a. They will lose weight.
   b. They will gain weight.
   c. They will neither lose nor gain weight.
   d. They will suffer from a lack of vitamins.

25. A good source of calcium is
   a. fruits.   c. meat.
   b. cottage cheese.   d. milk.

26. A diet low in iron can result in
   a. blindness.   c. scurvy.
   b. weight gain.   d. anemia.

27. A good source of Vitamin C is
   a. poultry.   c. oranges.
   b. cabbage.   d. tomatoes.

28. Iodine is important to maintain
   a. proper levels of thyroid hormones.
   b. proper insulin levels.
   c. healthy blood-sugar levels.
   d. bone cell development.

29. Calcium is important for
   a. strong bones.
   b. night vision.
   c. a healthy thyroid.
   d. strong teeth.

Name

Fitness & Health

> Professor Femur spends an hour exercising each morning before she goes to work.
> Before she starts this morning, she takes her pulse. It is 72 beats per minute.
> For the first 10 minutes, she stretches her shoulders, neck, arms, legs, and back.
> Then she walks on her treadmill for 10 minutes.
> Next, she does some weight-lifting for different parts of her body.
> Dr. Femur takes her pulse again. Now it is 90 bpm.
> Then she jumps in her swimming pool and swims laps for 15 minutes.
> She stops and checks her pulse. It is now 150 bpm.
> She continues swimming for another 20 minutes.
> When she dries off, she spends about 5 minutes stretching to finish her workout.

30. Which part of her workout is increasing her muscle flexibility? _____

31. Which part of her workout is increasing her muscle strength? _____

32. Which part of her workout is providing exercise to her heart and lungs? _____

33. What activities are warming up and cooling down her muscles? _____

34. Which of the following are benefits of regular exercise?
    a. It helps to maintain a healthy weight.
    b. It helps to relieve stress.
    c. It strengthens the heart.
    d. It helps produce substances to keep fatty deposits from building up in arteries.
    e. a, b, and c
    f. All of the above.

35. Which of these prevents the formation of fatty deposits in the walls of the arteries?
    a. high-density lipoproteins (HDLs)
    b. low-density lipoproteins (LDLs)

36. Which activities will provide your body with aerobic exercise?
    a. push-ups   d. power walking
    b. stretching   e. cross-country skiing
    c. jogging   f. weight-lifting

Fitness & Health

38. Which of the following is NOT a practice likely to help someone manage stress?
a. Increase calorie intake for several hours.
b. Lower intake of caffeine, salt, and sugar.
c. Increase intake of Vitamin C.
d. Exercise regularly.
e. Learn effective methods of solving problems.

37. Approximately how much water is good to drink on a daily basis?
a. 4 cups
b. 2 gallons
c. 2 liters
d. 10 liters

39. Which does NOT accurately describe a body response to stress?
a. Heart rate and blood pressure increase.
b. Adrenal glands increase production of adrenalin.
c. Extra sugar released in the blood stream gives the body extra energy.
d. Blood flow to the muscles is decreased.
e. Breathing speeds up; more breaths are taken per minute.

Match the emergency with a sensible **First Aid** response at the right.

A. SHOCK
B. FROSTBITE
C. SNAKEBITE
D. MINOR BURN
E. FAINTING
F. BAD FALL
G. HYPOTHERMIA
H. CHEMICAL BURN

_____ 40. Do not move the victim.
_____ 41. Rub ice gently on the area.
_____ 42. Wash the area with plenty of water
_____ 43. Keep the affected area below the heart level.
_____ 44. Place the affected area in slightly warm water or wrap in a blanket
_____ 45. Remove all damp clothing, keep the victim warm, give liquids.
_____ 46. Lower the head between the knees, or lie flat with legs raised.
_____ 47. Keep the victim lying down with feet slightly raised; cover with a blanket; give orange juice.

Name _____

Middle Grade Book of Science Tests

Circle the number of all statements that are true.

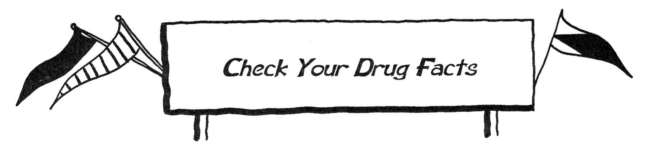

Check Your Drug Facts

48. Hallucinogens, such as LSD or PCP, do not produce psychological dependence.
49. Amphetamines increase heart rate, blood pressure, and rate of breathing.
50. Regular coffee drinking can cause higher heart rate and disturbed sleep.
51. Alcohol is absorbed more rapidly when the drinker has a full stomach.
52. The circulatory system and heart rate are affected by marijuana.
53. Nicotine in tobacco is a depressant. It slows body responses.
54. People can become psychologically dependent on caffeine.
55. Smoking can contribute to lung cancer and emphysema.
56. Tranquilizers can be purchased without a prescription.
57. Alcohol slows down the body processes and responses.
58. Use of cocaine sometimes leads to violent behavior.
59. Smokers become physically dependent on nicotine.
60. Crack is a name given to a strong form of cocaine.
61. Marijuana is neither a stimulant nor a depressant.
62. Cigarette smoke can be harmful to nonsmokers.
63. An overdose of barbiturates can cause death.
64. Long-term use of alcohol damages the liver.
65. Smoking can contribute to heart disease.
66. Tobacco use by anyone under 18 is illegal.
67. Most narcotics are not habit-forming.
68. Amphetamines are safe drugs to use.
69. Alcohol increases alertness.
70. All use of narcotics is illegal.

Physical Science Skills Checklists

Physical Science Test # 1 :
MATTER: STRUCTURE & PROPERTIES
Test Location: pages 94–97

Skill	*Test Items*
Distinguish between the three states of matter; identify characteristics of each (including viscosity)	1–2; 5–8, 14
Show understanding of the kinetic theory of matter	9
Recognize the effect of temperature on matter	10
Recognize the effect of pressure on a gas or liquid	11–12
Show understanding of the concept of buoyancy	13
Distinguish among solutions, mixtures, colloids, suspensions, homogeneous and heterogeneous mixtures	15–18
Distinguish between physical and chemical properties of a substance	19–30
Show understanding of the structure, components, and functions of the parts of the atom	31–35
Identify features of an atom from a model or description	36–38
Recognize the symbols for common elements	39–54
Identify features of an element from reading the periodic table	55–61
Show understanding of the groupings on the periodic table	62–65, 69–71
Identify some characteristics of elements	66–67
Distinguish between an element and a compound	68
Recognize organic compounds and hydrocarbons	69–70
Identify common compounds from their formulas	74–79
Write formulas for compounds shown in a diagram	80–85

Physical Science Test # 2 :
MATTER: CHANGES & INTERACTIONS
Test Location: pages 98–103

Skill	*Test Items*
Distinguish between physical and chemical changes in a substance	1–14
Show understanding of some characteristics of physical and chemical changes	15–16
Identify the physical changes of freezing, melting, evaporation, condensation; recognize freezing point, melting point, boiling point	17–27
Show understanding of processes and components in solutions	28–34
Identify solute, solvent, aqueous solutions, saturated, supersaturated, electrolyte	28–30, 34
Show understanding of how solubility of a substance can be increased	31–32
Demonstrate how a formula can be used to show the chemical reaction between two substances	35–37
Distinguish between balanced and unbalanced chemical equations	38
Show understanding of the characteristics and behavior of acids and bases	38–40

Physical Science Test # 3 :
FORCE & MOTION
Location: pages 104–109

Skill	Test Items
Identify and define force & motion concepts	1–10
Understand and calculate speed and velocity	11–13
Recognize details and applications of the laws of motion	14–19
Recognize and apply concepts of wind resistance, velocity, relative velocity, acceleration, terminal velocity, freefall, momentum, inertia	20–35

Physical Science Test # 4 :
ENERGY
Test Location: pages 110–117

Skill	Test Items
Define and distinguish between potential energy and kinetic energy	1–9
Determine amount of energy needed for heating task	9–10
Understand the concept of energy transfer	10, 13
Recognize and apply concepts of thermal energy, heat, and heat transfer	11–13, 41
Identify scientific units used to measure energy, force, work, power, and electric current	14–18
Describe the Law of Conservation of Energy	20
Define and distinguish among different kinds of energy	21–28
Recognize different methods of heat transfer	29–40
Define terms and processes related to work	42–46, 59, 60
Identify different simple machines	47–58
Define power, calculate mechanical advantage, work, and power	61–65, 87, 88
Define concepts, processes, and terms related to electricity and its use	66–73
Show understanding of workings of circuits	74–77, 80, 82–83
Explain static electricity	78
Show understanding of concepts relating to magnetism	79, 81, 83–86
Identify some radioactive elements	89
Explain nuclear fusion and nuclear fission	90

Physical Science Test # 5 :
WAVES, SOUND, LIGHT
Location: pages 118–121

Skill	Test Items
Recognize features of waves	1–10, 24
Show familiarity with the location of waves on the electromagnetic spectrum	11–13, 17–19
Show understanding of frequency and its relationship to wavelength	14–16
Use velocity to calculate distance of sound sources	20
Recognize the relationship between temperature and sound transfer	21
Show understanding of materials that conduct sound	22
Recognize relationship between amplitude and intensity of sound	23, 27
Show understanding of the Doppler effect of sound	25–26
Recognize and explain some properties of light	28–34
Distinguish between convex and concave lenses	35

Physical Science, Test #1

STRUCTURE & PROPERTIES OF MATTER

Name _____   Possible Correct Answers: 85

Date _____   Your Correct Answers: _____

1. A substance is poured from a small container into a large container. It fills the small container. It also fills the large container.
   The substance is a   a. solid.   b. liquid.   c. gas.

2. A substance has a definite shape and a definite volume.
   The substance is a   a. solid.   b. liquid.   c. gas.

3. All matter is made of tiny particles, called _____.

4. Matter composed of one kind of matter is an _____.

5. All matter has _____ and _____.

6. Professor Reginald Radon examines these substances. Which is a liquid? _____

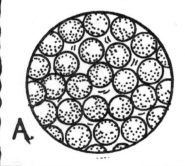
A.

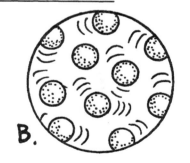
B.

C.

7. The state of matter of most substances can be changed. At any given time, the state a substance is in depends on its
   a. size.   c. temperature.
   b. viscosity.   d. color.

8. Which form of matter contracts and expands the most?
   a. solids
   b. liquids
   c. gases

9. Which statement is NOT a part of the kinetic theory of matter?
   a. All matter can change.
   b. Particles in matter are in constant motion.
   c. All matter is made of tiny particles.
   d. The motion and spacing of the particles determines the state of matter.

10. A pot of broth is set on a stove. Which of these will happen as the temperature of the broth rises?
    (*Choose one or more answers.*)
    a. The particles will move faster.
    b. The volume of the broth will increase.
    c. The particles will collide with each other more violently.
    d. The space between the particles will diminish.

11. Dr. Radon blows up a balloon. He ties it tightly, and squeezes the balloon to make it smaller. What will happen to the pressure of the air inside?
    a. The pressure of the air inside will increase.
    b. The pressure of the air inside will decrease.
    c. The pressure will remain the same.

12. The professor squeezes a container of orange juice, putting pressure on the outside edges of the container. The pressure on the juice in the center of the container will be
    a. greater than on the outside edges.
    b. less than on the outside edges.
    c. the same as on the outside edges.

13. A wooden block is placed in a bathtub full of water. The buoyant force exerted on the block is less than the force of gravity. The block will
    a. sink
    b. float

14. Which substance has the highest viscosity?
    a. water
    b. maple syrup
    c. tar
    d. chocolate milk

Matter: Structure & Properties

15. Abigail brews up a pot of tea. She pours it over ice until all the ice is melted. She is drinking a
   a. solution.
   b. homogenous mixture.
   c. colloid.
   d. suspension.

16. Alfredo mixes up a chocolate milkshake. It has smooth creamy liquid, icy chunks of ice cream, and crumbles of chocolate chips. This substance is a
   a. solution.
   b. homogeneous mixture.
   c. colloid.
   d. suspension.

17. Antonio is examining the properties of whipped cream, smog, mayonnaise, and soapsuds. These are all
   a. heterogeneous mixtures.
   b. colloids.
   c. solutions.

18. Andrea pours some tangy salad dressing on her salad. When she sets the bottle down, the liquids in the salad dressing begin to separate. The spices settle to the bottom. Her salad dressing is a
   a. homogeneous mixture.
   b. heterogeneous mixture.
   c. solution.

Matter has physical properties and chemical properties. Label each property below P (for physical) or C (for chemical).

____ 19. color
____ 20. corrosion
____ 21. temperature
____ 22. smell
____ 23. hardness
____ 24. magnetism

____ 25. its shape
____ 26. density
____ 27. how it burns
____ 28. its boiling point
____ 29. its ability to conduct electricity
____ 30. how it reacts with other substances

Name _____

31. What part of the atom holds most of the mass of the atom? _____

32. Which part of an atom carries no electric charge? _____

33. Which part of an atom carries a positive charge? _____

34. Which part of an atom carries a negative charge? _____

35. How many electrons can be held in the first orbit in an atom? _____

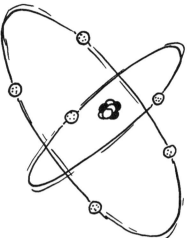

Each diagram below represents the atom of an element.

36. What is the atomic number of this element? _____

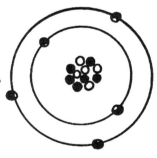

37. What is the atomic mass of this element? _____

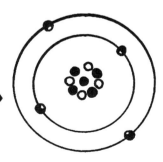

38. An atom has an atomic number of 19.
    Its atomic mass is 39.
    How many neutrons does it have? _____

Matter: Structure & Properties

Match the elements with their symbols.
Write the symbol.

____39. copper

____40. sodium

____41. lead

____42. oxygen

____43. neon

____44. mercury

____45. chlorine

____46. hydrogen

K Pb C
O Na Mg
Ne Hg
    Au
Cu Ag
Al   H
    N
Fe   Cl

____47. gold

____48. silver

____49. carbon

____50. aluminum

____51. iron

____52. nitrogen

____53. potassium

____54. magnesium

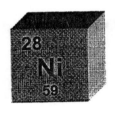

____55. What is the atomic number of this element?

____56. How many neutrons does it have?

____57. What is its atomic mass?

____58. What is the atomic mass of this element?

____59. How many electrons does it have?

____60. How many protons does it have?

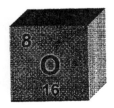

____61. How many electrons will this atom have in the second energy level?

62. Which of the following elements are NOT in the halogen family?

fluorine   iodine   phosphorus   chlorine   bromine   tin

63. Which of the following elements are in the nonmetal group on the periodic table?

neon   krypton   carbon   barium   nitrogen   lithium

64. Which of the following are metallic elements on the periodic table?

potassium   calcium   lead   radium   selenium   silicon   carbon

65. Which of the following are transition elements on the periodic table?

hydrogen   iron   copper   gold   sodium   arsenic   nickel   neon

66. A mineral or other natural material from which metal can be obtained is
   a. an alloy.   c. an ore.
   b. a ductile.   d. an alkaline element.

67. A child has been warned never to touch an element. This is the only metal that is liquid at room temperature.

   What is this element? _____

68. Explain the difference between an element and a compound.

_____
_____
_____
_____
_____

Matter: Structure & Properties

69. Dr. Radon is examining an element.
It is very hard, and has a high melting point.
It is very highly-colored.
It has two electrons in its outer energy level.

On the periodic table, it can probably be found
a. on the left side, with the metallic elements.
b. in the center, with the transition elements.
c. on the right side, with the nonmetallic elements

70. Next, Dr. Radon examines a shiny element.
It has three electrons in the outer energy level.
It conducts heat and electricity well.
It is very hard.
It is malleable (can be beaten into a flat sheet).

On the periodic table, it can probably be found
a. on the left side, with the metallic elements.
b. in the center, with the transition elements.
c. on the right side, with the nonmetallic elements.

71. Dr. Radon examines a third element.
It is a good insulator.
It does not conduct heat or electricity well.
It is not at all shiny.
It cannot be flattened into a sheet
It cannot be pulled into a wire.
It tends to gain or share electrons when it reacts with other substance.

On the periodic table, it can probably be found
a. on the left side, with the metallic elements.
b. in the center, with the transition elements.
c. on the right side, with the nonmetallic elements.

72. Which of the compounds shown below is NOT an organic compound? _____

73. Which compounds below are hydrocarbons? _____

$AgNO_3$ C_7H_{16} CO_2

CH CH_4

$NaCHO_3$

$CaCO_3$

Write the name of these common compounds.

74. H_2O_2 _____

75. SiO_2 _____

76. HCl _____

77. CO_2 _____

78. CO _____

79. KBr _____

Write a formula for each compound shown.

80. _____

81. _____

82. _____

83. _____

84. _____

85. _____

Physical Science, Test #2

MATTER: CHANGES & INTERACTIONS

Name _____   Possible Correct Answers: 40

Date _____   Your Correct Answers: _____

Things are changing all around Professor Radon's world.
Some changes are physical changes.
Others are chemical changes.
The professor knows the difference. Do you?
Label each change **P** (physical) or **C** (chemical).

_____ 1. This morning, Dr. Radon notices that his wife bleached her hair.

_____ 2. Leaving the house, he sees that his son's bicycle is rusting badly.

_____ 3. At the lab, Dr. Radon dropped a glass beaker. It shatters.

_____ 4. He builds a fire in the fireplace. Soon the wood is crackling as it burns.

_____ 5. The fica plant in the corner of the lab is busy photosynthesizing.

_____ 6. The wax that melted in an experiment is now hardening.

_____ 7. Dr. Radon uses a sieve to separate the sand from gravel in a dirt mixture.

_____ 8. Dr. Randon mixes up a potion of salt water to gargle for his sore throat.

_____ 9. Ooops! His lunch was left out of the refrigerator. The sandwich has spoiled.

_____ 10. Since the sandwich is inedible, the professor fries an egg for lunch.

_____ 11. He also dissolves some powder in milk to make hot chocolate.

_____ 12. The steamy hot chocolate leads to a misty fog on the windows of the lab.

_____ 13. Absent-minded Professor Radon forgot to watch his toast. It has burned black.

_____ 14. The professor's lunch is digesting in his stomach.

Matter: Changes & Interactions

15. In a chemical change what could be released?
    a. heat
    b. electricity
    c. light
    d. a and c
    e. all of the above

16. A physical change could change any of the following, except
    a. color.
    b. shape.
    c. state of matter.
    d. chemical make-up.
    e. size.

Read the events described A–H. Write a matching letter next to each term or label listed in items 17–27. A letter may be used more than once. Some answers will use more than one letter.

_____ 17. evaporation

_____ 18. melting point

_____ 19. 32° F

_____ 20. freezing point

_____ 21. 212° F

_____ 22. condensation

_____ 23. freezing

_____ 24. boiling point

_____ 25. 0° C

_____ 26. melting

_____ 27. 100° C

A. The water in the birdbath just turned to ice.

B. Finally, the ice is melting off the windshield.

C. The grape juice you poured into little trays and put into the freezer last night has turned into tasty grape popsicles.

D. Yesterday, your little sister built a snowman. Today, it is just a pile of slush.

E. You boiled some water to cook pasta. You forgot to put in the pasta. Now all the water has disappeared.

F. On a cold winter day, the greenhouse is warm and damp. Little beads of water form on the insides of all the greenhouse windows.

G. You left a damp sponge on the counter last night. This morning, it is completely dry.

H. Your water has just begun to bubble and turn into water vapor.

Matter: Changes & Interactions

Professor Radon's Solutions
Professor Reginald Radon is mixing up these solutions.

A. He stirs sugar and lemon juice into water to make lemonade.
B. He uses a special container to mix carbon dioxide and cherry juice to make cherry soda.
C. He stirs a chocolate powder and sugar cubes into milk to make a yummy drink.
D. He makes some antifreeze by mixing ethylene glycol and water.
E. He dissolves salt in water.
F. He combines alcohol and water.
G. He dissolves potassium bromide in water.

28. Which substances are solutes? _____

29. Which substances are solvents? _____

30. Which solutions are aqueous solutions? _____

31. The sugar-chocolate mixture is not dissolving well in the milk. What can the professor do to the ingredients to help solve this problem?
(Choose one or more answers.)

  a. Shake or stir them.   b. Cool the milk.   c. Heat the milk.   d. Crush the sugar cubes.

32. What can the professor do to dissolve more carbon dioxide in the cherry juice?
  a. Heat the cherry juice.
  b. Cool the cherry juice.
  c. Nothing can help him do this.

33. He keeps adding salt to the water. Some of the salt crystals settle out when he stops stirring. This happens because the solution is
  a. aqueous.
  b. saturated.
  c. supersaturated.

34. The potassium bromide solution acts as a conductor of electricity. The alcohol solution will not conduct electricity. Which substance is an electrolyte?

  _____

Name _____

35. Write a formula to show this reaction:
    carbon plus oxygen yields carbon dioxide: _____

36. Write a formula to show this reaction:
    barium chloride decomposes into barium and chlorine gas. _____

37. Which equations are balanced?

   a. $H_2O \longrightarrow 2H + O$

   b. $2KClO_3 \longrightarrow 2KCl + 3O_2$

   c. $2Mg + O_2 \longrightarrow 2MgO$

   d. $Mg + O_2 \longrightarrow MgO$

Professor Radon is dissolving several substances into water. Here are some characteristics of his solutions.

Solution A produces hydroxide ions.
Solution B has a bitter taste.
Solution C has a sour taste.
When phenolphthalein is added to **solution D**, it turns pink.
Solution E is slippery.
Solution F reacts with metal.
Solution G produces hydronium ions.
Solution H has a pH scale reading of 7.
Solution I has a pH scale reading of 8.
Blue litmus paper placed in **solution J** turns red.
Red litmus paper placed in **solution K** turns blue.
Litmus paper placed in **solution L** shows no change.

38. Which solutions are acids? _____

39. Which solutions are bases? _____

40. Which solutions are neutral? _____

Physical Science, Test #3

FORCE & MOTION

Name _____

Date _____

Possible Correct Answers: 35

Your Correct Answers: _____

The student's definitions of force & motion concepts are terribly confused. Straighten out the mess. Draw a line to the correct definition for each concept.

1. **gravity**
2. **acceleration**
3. **force**
4. **friction**
5. **velocity**
6. **momentum**
7. **inertia**
8. **speed**
9. **mass**
10. **freefall**

a. speed and direction of a moving object
b. force that objects exert on each other
c. rate of change in the position of an object
d. product of an object's mass and velocity
e. a push or pull on an object
f. rate of change in velocity
g. the amount of matter in a body
h. the property of a body that resists change in movement
i. force opposing motion between surfaces touching each other
j. unrestrained motion in place where gravity is acting

11. A train travels towards the west 1020 miles in 12 hours. What is the train's speed?

_____

12. Two trains are 2000 miles apart on the same track. One travels east at a velocity of 150 mph. The other travels west at a velocity of 100 mph. How long will it take for them to meet *(assuming there are no stops)*?

_____

13. Two trains travel at 120 mph for 7 ½ hours. One train travels east. The other travels west. Do the trains have the same velocity?

(Circle one answer.)

YES   NO

LAWS of MOTION

Law # 1 Every object remains still or continues moving at a steady rate in a straight line unless acted upon by another force. *(This is the inertia principle.)*

Law # 2 The amount of force needed to change the speed of an object depends on the object's mass and the amount of acceleration. Acceleration of an object increases as the force applied increases.

Law # 3 For every action there is an equal and opposite reaction. *(When an object is pushed one way by a force, an equal force pushes the other way.)*

For each example below, write the number of the law that applies.

_____ 14. A small child pitches a baseball. A professional pitcher pitches a baseball of the same size and weight. The pitcher's ball travels faster and farther.

_____ 15. A girl is standing on a bus that is stopped at a stoplight. The bus lurches forward. The upper part of her body is jerked backward.

_____ 16. The jet fuel in a rocket burns and produces a powerful blast out the back of the rocket. The rocket zooms forward.

_____ 17. A little girl is pulling a sled. She stops, but the sled keeps moving and slams into the back of her legs.

_____ 18. A pair of campers paddles a canoe across a lake. They dig their paddles into the water and pull them backward. The canoe moves forward.

_____ 19. Members of a bobsled team are not satisfied with the speed of their bobsled starts. Each member spends six months at a gym working on building muscle. Six months later, they find that they can accelerate their sled much more quickly.

Force & Motion

20. Professor Radon drops a feather out the window of his laboratory. At first the feather falls quickly, then it slows down and falls very slowly. This is because
a. the feather is so light that gravity has little pulling on it.
b. air resistance pushes against the feather and slows its speed.
c. centripetal force pushes the feather in a circular path.

A bus is traveling along a road at a velocity of 45 kilometers per hour. A woman on the bus walks from the rear of the bus toward the front at a velocity of 3 kilometers per hour.

21. Someone on the bus would measure her speed at
a. 45 km/h.   b. 42 km/h.   c. 0 km/h.   d. 3 km/h.   e. 48 km/h.

22. Someone standing along the road would measure her speed at
a. 45 km/h.   b. 42 km/h.   c. 0 km/h.   d. 3 km/h.   e. 48 km/h.

23. A motorcycle accelerates from 5 mph to 35 mph in 15 seconds. What is its rate of acceleration?
a. 2 miles per second^2
b. 30 miles per second^2
c. 3 miles per second^2
d. 0.5 miles per second^2

24. Which object, when falling, will have a lower terminal velocity?
a. ping pong ball
b. tennis ball
c. pool ball
d. 10-pound rock

25. A 150-pound man steps into an elevator. Suddenly, the elevator rises at a rapid rate, speeding from the first to the 14th floor. If the man stepped onto a scale in the elevator during the rapid rise, he would weigh
a. 150 pounds.
b. more than 150 pounds.
c. less than 150 pounds.

Force & Motion

26. A bowler rolls a bowling ball. At the same time, a golfer hits a golf ball and a ping-pong player hits a ping pong ball.
If the three balls have the same momentum, which ball has the greater velocity?

   a. the bowling ball
   b. the golf ball
   c. the ping-pong ball

27. A small child and a strong adult each throw a tennis ball. Both of them throw the ball as hard as they possibly can. The two balls have the same size and weight. Which is true?

   a. Both balls will have the same momentum.
   b. Both balls will have the same velocity.
   c. Both balls will have the same amount of inertia.

Choose one of these words to complete each sentence. A word may be used more than once.

- GRAVITY
- MOMENTUM
- INERTIA
- CENTRIPETAL FORCE
- VELOCITY
- AIR RESISTANCE
- FRICTION
- ACCELERATION

28. _____ keeps Max from falling out of an upside-down roller-coaster car.

29. A space capsule entering Earth's atmosphere glows red-hot because of _____ .

30. _____ slows the fall of a leaf that has just been blown off a tree.

31. Jan could get her bike moving from 0 to 12 mph in 30 seconds. Now her rate of _____ has increased, and she can get from 0 to 12 mph in 25 seconds.

32. J.R. shoots an arrow toward a target. The arrow is moving straight for the target, but _____ pulls it to the ground before it reaches the target.

33. A pool player slams a red ball into a white ball. The red ball stops and the white ball moves. The _____ of the red ball was transferred to the white ball.

34. A ski racer loses a ski when he falls at the bottom of the hill. His body stops moving, but the ski keeps on going because of _____ .

35. Andy pushes his little brother's sled along a flat, snowy path. Eventually the sled is stopped by _____ .

Name _____

Physical Science, Test #4

ENERGY

Name _____

Date _____

Possible Correct Answers: 90

Your Correct Answers: _____

Write **P** for examples of potential energy.
Write **K** for examples of kinetic energy.

_____ 1. moving fan
_____ 2. rolling bowling ball
_____ 3. a stretched rubber band
_____ 4. a speeding train
_____ 5. a swinging swing
_____ 6. a charged battery
_____ 7. gasoline pump that is NOT running

8. Which example shows **kinetic energy**?
   A   B

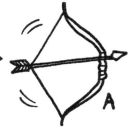

9. At which point does the roller coaster show **potential energy**?
   A   B   C

10. Which kind of energy transfer happens when the coaster moves from point B to point C?
    a. potential energy to kinetic energy
    b. kinetic energy to potential energy

11. Dr. Radon is measuring the average kinetic energy of the particles in a substance. What is he trying to find out about the substance?
    a. its weight   b. its state of motion   c. its temperature   d. its velocity

12. When his hands are cold, he rubs them together. What will happen?
    a. The movement in his hands will raise his body temperature.
    b. The kinetic energy of the moving hands will change into thermal energy.
    c. The kinetic energy of the moving hands will change to potential energy.

13. Heat always transfers
    a. from an object at a high temperature to an object at a lower temperature.
    b. from an object at a low temperature to an object at a higher temperature.
    c. between objects of the same temperature.

Write the correct unit.
Which unit measures ...

_____ 14. ... energy?

_____ 15. ... force?

_____ 16. ... power?

_____ 17. ... electric current?

_____ 18. ... work?

newtons *amperes* *watts* *joules*

The specific heat of water is 4190 J(joules)
(This is the amount of energy needed to raise 1 kg of water 1° C.)

19. Dr. Radon gets ready to heat a pot of water. The pot holds 2 kg of water. The water is 10° C. He will heat the water to 60° C. How much energy will be used?

    a. 20 J   c. 209,500 J
    b. 8380 J   d. 419,000 J

20. Explain the Law of Conservation of Energy.

Energy

Kind of Energy	Definition/Description	Example or Use
A.	energy of motion	moving train
potential energy	energy an object has as a result of its position or condition	B.
work	C.	pushing a suitcase
D.	total energy of the particles that make up a body	hands rubbing together
electric energy	E.	light bulb
radiant energy	F.	treatments used to kill diseased cells
G.	energy created by attractions of force in magnetic fields	loudspeakers
nuclear energy	H.	production of electricity

Each label below is missing from its correct spot on the chart.
Write the letter that shows where each one belongs in the chart.

_____ 21. power created when electrons and other charged particles flow through a conductor

_____ 22. energy produced by the combining or splitting of atoms

_____ 23. transfer of energy as a result of motion

_____ 24. energy that travels through space in the form of waves

_____ 25. a bent pole-vault pole

_____ 26. magnetic energy

_____ 27. thermal energy

_____ 28. kinetic energy

Label each example or description with the matching form of heat transfer. Choose one or more of these.

A. conduction B. convection C. radiation

_____ 29. The sun warms your face.

_____ 30. You warm up your soup in a pan on the stove.

_____ 31. Ocean currents warm colder water.

_____ 32. Thermal energy is transferred without matter.

_____ 33. A forced-air furnace blows heated air around the house.

_____ 34. You place your cold hands under warm water.

_____ 35. Heat is transferred by the movement of matter.

_____ 36. You stir hot stew with a spoon. The handle of the metal spoon gets hot.

_____ 37. Your feet burn when you walk barefoot across hot sand.

_____ 38. Thermal energy is transferred through matter from particle to particle.

_____ 39. Two objects at different temperatures are in direct contact.

40. Which is an explanation for the movement of air and water in currents?
   a. When air or water is heated, its density decreases and the air or water rises.
   b. When air or water is heated, its density increases, and the air or water sinks.
   c. When air or water is heated, its density decreases and the air or water sinks.
   d. When air or water is heated, its density decreases, and the air or water sinks.

41. Which of the following substances are poor conductors of thermal energy?
   wood   glass   copper   plastic   iron   fiberglass   air

Energy

Define each of these:

42. **work** _____

43. **mechanical advantage** _____

44. **simple machine** _____

45. **effort force** _____

46. **resistance force** _____

Name the simple machine that matches each description.

_____ 47. A bar pivots on a fixed point.

_____ 48. This device changes the direction of the force.

_____ 49. A large wheel and a small wheel rotate together.

_____ 50. This device is an inclined plane wound around a cylinder.

_____ 51. This device is an inclined plane with one or two sloping sides

_____ 52. A slanted surface allows movement of heavy objects by sliding them

53–58. Name the simple machine pictured in each example.

53. _____   54. _____   55. _____

56. _____   57. _____   58. _____

Name _____

59. In this diagram, which point represents the resistance force?

   A   B   C

60. What is the mechanical advantage of the screwdriver?
   a. 11   c. 44
   b. 4.4   d. 110

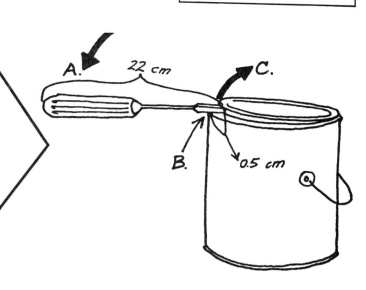

61. A hammer with a mechanical advantage of 15 is used to pull out a nail that has a resistance force of 4500 newtons. What effort force is needed to pull the nail out of the wood beam?
   a. 1085 N
   b. 33.3 N
   c. 300 N

62. A child pushes a box of toys across the floor. The force used is 10 newtons. She pushes the box 25 meters. How much work is done?
   a. 25 J
   b. 250 J
   c. 35 J

63. 285 J of work is done when a piano is lifted. The force used was 95 newtons. What distance was the piano lifted (in meters)?
   a. 380 m
   b. 3 m
   c. 27,975 m

64. Power is
   a. the amount of work done.
   b. the product of effort and force.
   c. the amount of work done per unit of time.
   d. the amount of work done per unit of force.

65. An elevator weighs 30,000 N. It is lifted 30 m in 50 seconds. How much power is needed to do this job?
   a. 18,000 W   c. 50,000 W
   b. 90,000 W   d. 500,000 W

Energy

page 7 of 8

Circle the items that the student has done correctly.
If an item is incorrect, cross out the wrong answer, and write the correct answer.

ELECTRICITY QUIZ
Match the names with their descriptions.

____ _H_ 66. *electric charge*   **A.** material through which electrons can move

____ _G_ 67. *static electricity*   **B.** flow of electrons through a conductor

____ _D_ 68. *insulator*   **C.** a device to break a circuit

____ _E_ 69. *resistance*   **D.** a poor conductor of electricity

____ _F_ 70. *circuit*   **E.** having too many or too few electrons

____ _C_ 71. *conductor*   **F.** path of electric conductors

____ _B_ 72. *electric current*   **G.** electric charge built up in one place

____ _A_ 73. *switch*   **H.** opposition to the flow of electricity

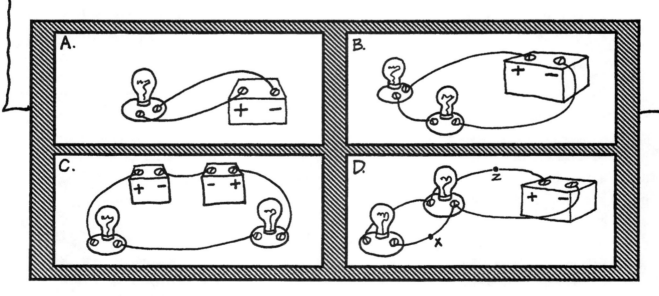

_____ 74. Which pictures above show series circuits?

_____ 75. In which picture will the bulbs NOT light?

_____ 76. If circuit D is broken at point X, how many bulbs will light?

_____ 77. If circuit D is broken at point Z, how many bulbs will light?

Name _____

Energy

78. Explain why Professor Radon's hair is standing up when he combs it with a plastic comb on a dry day.

_____
_____
_____
_____

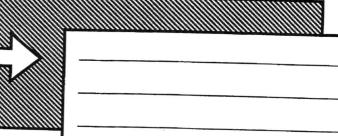

Circle the numbers of the statements that are true.

79. Like magnetic poles attract.

80. A direct current changes directions.

81. An electric charge creates a magnetic field.

82. Electrons move easily through an insulator.

83. Electromagnetic energy travels by convection.

84. A magnetic field is strongest around the ends of a magnet.

85. A transformer is a device that causes a material to be magnetic.

86. A magnet is held over a pile of coins (pennies, nickels, dimes, quarters). Which coin is most likely to be attracted to the magnet?

_____

87. A hairdryer uses 1800 watts of power from a source of 120 volts. What is the current?
 a. 150 A
 b. 15 A
 c. 226,000 A

88. A 120-volt source of power is used to cook a roast beef in an oven. The current is 20A. How much power is delivered to the oven?
 a. 2400 V
 b. 60 V
 c. 6 V

89. Which of these elements exist naturally as radioactive elements?

uranium   radium   radon   barium
   neon   polonium

90. Explain the difference between nuclear fusion and nuclear fission.

_____
_____

Name _____

Physical Science, Test #5

WAVES, SOUND, & LIGHT

Name _____

Date _____

Possible Correct Answers: 35

Your Correct Answers: _____

What do you know about waves?
*Match the right word with its description.
Refer to the wave diagram.*

_____ 1. the speed and direction of the wave

_____ 2. a unit of wave frequency equal to one cycle per second

_____ 3. the transfer of energy by electromagnetic waves

_____ 4. the number of waves that pass point D per second

_____ 5. the part of the wave shown at letter A

_____ 6. the rapid back and forth movement of wave particles

_____ 7. the wave feature shown by arrow C

_____ 8. the kind of wave shown here

_____ 9. the part of the wave shown at letter B

_____ 10. the wave feature shown by arrow F

A. amplitude
B. wavelength
C. trough
D. frequency
E. transverse
F. radiation
G. vibrations
H. hertz
I. velocity
J. crest

Middle Grade Book of Science Tests

Copyright ©2001 by Incentive Publications, Inc., Nashville, TN.

11. Which waves on the electromagnetic spectrum are visible?
   a. radar waves
   b. ultraviolet rays
   c. light waves
   d. radio waves

12. Of these waves, which would have the higher frequency?
   a. radio waves
   b. infrared rays
   c. gamma rays
   d. TV waves

13. Of these waves, which would have the shorter wavelength?
   a. cosmic waves
   b. visible light waves
   c. TV waves
   d. microwaves

14. A wave passes a point 80 times per second. What is its frequency?

15. As a wave length is shortened,
   a. the frequency of the wave is higher.
   b. the frequency of the wave is lower.

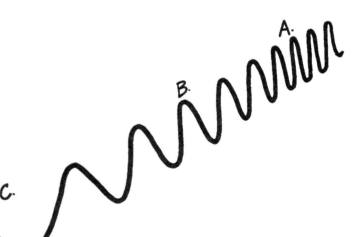

16. Wave A has a frequency of 56 Hz. Wave B has a frequency of 10 MHz. Which wave has the longer wavelength?

17. Which letter on the wave would show the location of visible light? _____

18. Which letter would show the location of radio waves? _____

19. Which letter would show the location of X-rays? _____

Waves, Sound, & Light

20. Tonight Professor Radon is listening to a thunderstorm outside the lab. The velocity of sound through air tonight is 345 m/s (meters per second). He hears a crash of thunder 4 seconds after seeing the flash of lightning that caused it. How far away from the lab is the storm?

_____

21. If the temperature were warmer,
    a. Dr. Radon would hear the thunder sooner.
    b. Dr. Radon would hear the thunder later.
    c. Dr. Radon would still hear the thunder in 4 seconds.

22. A banging sound would travel faster through
    a. air   b. water

23. The sound of the thunder is very loud. The sound wave has
    a. a large amplitude.   b. a small amplitude.

24. These two sound waves differ in
    a. amplitude.   c. all three.
    b. frequency.   d. none of these.
    c. wavelength.

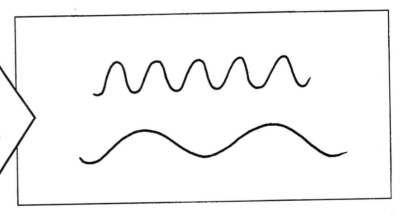

The lawnmower causes sound waves.

The lawnmower causes sound waves.

_____ 25. At what point is the frequency of the waves lower?

_____ 26. Does the sound have a higher pitch at point A or point B?

27. Which sound would have a loudness of about 100 dB (decibels)?
   a. a whisper   b. a power lawnmower   c. a purring cat   d. a jet plane taking off

28. What property of light is shown here? _____

29. What property of light is shown here? _____

30. The color of unseparated light is _____

31. Material that absorbs light is _____

32. The colored bands of visible light are known as the _____

33. A material that you can see through is _____

34. An apple is red because _____

35. What kind of the lenses described is a convex lens?   A   B

A. It is thicker in the middle than at the edges. It bends light rays together to a focal point. It can form a real image because it can focus light. It makes an object appear larger.

B. It is thinner in the middle than at the edges. It bends light rays away from each other. It cannot form a real image because it does not focus light.

Earth & Space Science Skills Checklists

Earth & Space Science Test # 1 :
SPACE
Test Location: pages 125–129

Skill	Test Items
Recognize features of the solar system	1–6
Recognize features and function of the sun	7–16
Identify features and position of the nine planets	17–28
Distinguish between different motions of planets and bodies in space	29–33
Recognize how revolution of Earth leads to different seasons	34–36
Recognize phases of the moon	37–43
Differentiate between a solar eclipse and a lunar eclipse	44–45
Label diagrams to explain what happens in eclipses	46–54
Identify bodies and features of outer space	55–67
Recognize and identify features and behaviors of comets	57, 68–69
Recognize some key persons, events, and missions in space exploration	70–75

Earth & Space Science Test # 2 :
EARTH COMPOSITION
Test Location: pages 130–133

Skill	Test Items
Recognize and define surface features on the Earth	1–4
Recognize the concept of relief and the use for topographic maps	5–6
Understand the use of latitude and longitude as a system for locating places on Earth's surface	7–8
Recognize features of Earth's surface and interior structure	9–17
Understand the rock cycle. Show understanding of the processes by which rocks are formed	18–22
Identify characteristics of different rocks	23–34
Identify the three main groups of rocks and how they are formed	26, 27, 31
Distinguish between igneous, metamorphic, and sedimentary rocks; identify specific rocks in these categories	35–44
Show understanding of the classification of minerals by hardness	45
Use understanding of mineral characteristics and classification to identify specific minerals	46–50

Earth & Space Science Test # 3:
EARTH CHANGES & PROCESSES
Test Location: pages 134–139

Skill   *Test Items*

Skill	Test Items
Distinguish between internal and external processes of change	1
Distinguish between disintegration and decomposition as two kinds of weathering	2–3
Identify ways gravity erodes the land	4
Distinguish between fast and slow movements of land	5
Name the agents of erosion	6
Explain the difference between weathering and erosion	7
Recognize features and processes of wind erosion	8–14
Recognize features and processes of erosion by moving water and groundwater	15–26
Recognize features and processes of glacial erosion	27–38
Identify different kinds of volcanoes	39–44
Identify features of volcanic activity	45–47
Show understanding of plate tectonics and earthquake processes	48–60
Identify features of earthquake activity	56–60

Earth & Space Science Test # 4:
RIVERS, STREAMS, LAKES, & GROUNDWATER
Test Location: pages 140–143

Skill	Test Items
Recognize processes by which moving water changes land	1–5
Identify and define features of river systems and runoff	6–12
Identify different kinds of river deposits	11, 12, 14
Recognize river drainage patterns	13
Recognize features of lakes	15, 16
Recognize features of a swamp or bog	17
Show understanding of the processes by which groundwater changes the land	18–25
Identify specific features related to groundwater	19–25

Earth & Space Science Test # 5 :
OCEANS
Test Location: pages 144–147

Skill	Test Items
Show understanding of general information about Earth's lithosphere	1
Show understanding of formation and processes of ocean currents	2–11
Recognize causes of waves	12–13
Identify features of waves	14–23
Show understanding of processes that occur as waves move toward shore	17
Distinguish between spring tides and neap tides	24–27, 31–33
Show understanding of the relationship between the sun-moon-Earth system and the movement of tides	28–33
Identify some features of shore zones	34–39
Identify some conditions necessary for the growth of coral reefs; distinguish among kinds of coral reefs	36, 37
Identify features of the ocean floor	40–45

Earth & Space Science Test # 6 :
AIR & WEATHER
Test Location: pages 148–152

Skill	Test Items
Identify layers of Earth's atmosphere and their features	1–8
Show understanding of atmospheric pressure	9
Discriminate among different kinds of winds and other air movements	10–18
Identify different weather conditions	19–31
Identify different weather fronts	32–38
Recognize features of different climate zones	39–45

Earth & Space Science, Test #1

SPACE

Name _____   Possible Correct Answers: 75

Date _____   Your Correct Answers: _____

Professor Agnes Igneous has made some notes in her log about the solar system. Find the missing words. Write the letter of the word that belongs in each blank.

- A. solar flares
- B. elliptical
- C. perihelion
- D. chromosphere
- E. photosphere
- F. revolve
- G. core
- H. oxygen
- I. circular
- J. radiant
- K. rotate
- L. aphelion
- M. hydrogen
- N. eclipse
- O. orbits
- P. corona
- Q. aurora borealis
- R. sunspots
- S. Earth
- T. sun

Before the discoveries of Copernicus, it was believed that 1 **S** was the center of the solar system. Since that time, it has been learned that Earth and other planets 2 **F** around the sun in paths called 3 **O** that are 4 **B** in shape. The point at which a planet is closest to the sun is its 5 **C** . The point at which it is farthest from the sun is its 6 **L** . The sun is a star. Most of its mass is 7 **M** and helium. At the 8 **G** of the sun, fusion occurs, sending 9 **J** energy outward. The 10 **E** is the surface of the sun. Beyond this is the 11 **D** , a bright red layer extending far above the surface. Above this is a transparent zone called the 12 **P** , which is only visible during a total 13 **N** of the sun. 14 **R** are relatively cool, dark areas on the sun's surface. Sudden increases in brightness of the chromosphere, called 15 **A** , are caused when electrons and protons stream outward at high speeds. When the gases in Earth's upper atmosphere are disturbed by solar flares, they cause a light show known in the Northern Hemisphere as 16 **Q** .

Space

Write the name of the planet that matches each description.

Mercury 17. closest to the sun
Jupiter 18. has the largest moon
Pluto 19. the smallest planet
Venus 20. similar size as Earth
Saturn 21. over 1000 rings

Mar 22. appears red due to red dust
Venus 23. has a retrograde rotation
Uranus 24. 10 rings and 15 moons
Earth 25. 3rd planet from the sun

Neptune 26. 8th from sun
Pluto 27. polar ice caps
Jupiter 28. the largest planet

WHICH PLANET?
Earth   Pluto
Jupiter Saturn
Mars   Uranus
Mercury Venus
Neptune

WHICH MOVES?
retrograde motion
real motion
apparent motion
rotation
revolution

Which kind of movement is described? Write the movement for each question.

rotation 29. Neptune turns on its axis.
revoluto 30. The Earth orbits around the sun.
apparent 31. To early observers it appeared that the sun circled the Earth.
revolution 32. It takes Saturn 29.5 Earth years to travel around the sun.
retrograde 33. One planet rotates the opposite direction from the others.

Space

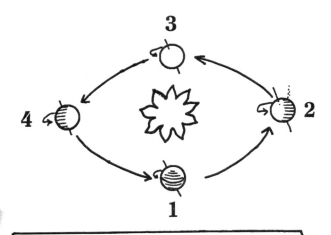

Write a number from the diagram.

__4__ 34. Which shows summer solstice in the Northern Hemisphere?

__3__ 35. Which shows fall equinox in the Southern Hemisphere?

__4__ 36. Which shows winter solstice in the Southern Hemisphere?

Write the letter to label each moon phase.

__C__ 37. the new moon
__A__ 38. first quarter
__B__ 39. waxing crescent
__E__ 40. third quarter
__H__ 41. waxing gibbous
__G__ 42. full moon
__F__ 43. waning gibbous

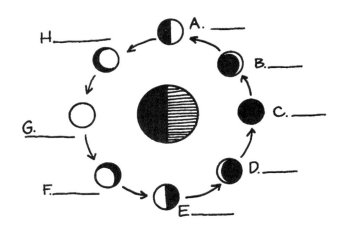

44. What kind of an eclipse is shown below? __lunar eclipse__

45. Which body is not visible during this eclipse? __umbra__

46. Explain why the body is not visible
__because the moon is blocking the sun light from it.__

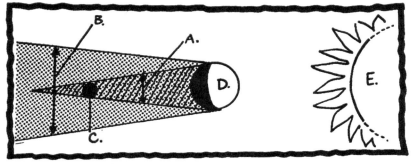

47. The moon is shown by letter __C__.

48. The umbra is shown by letter __A__.

49. The penumbra is shown by letter __B__.

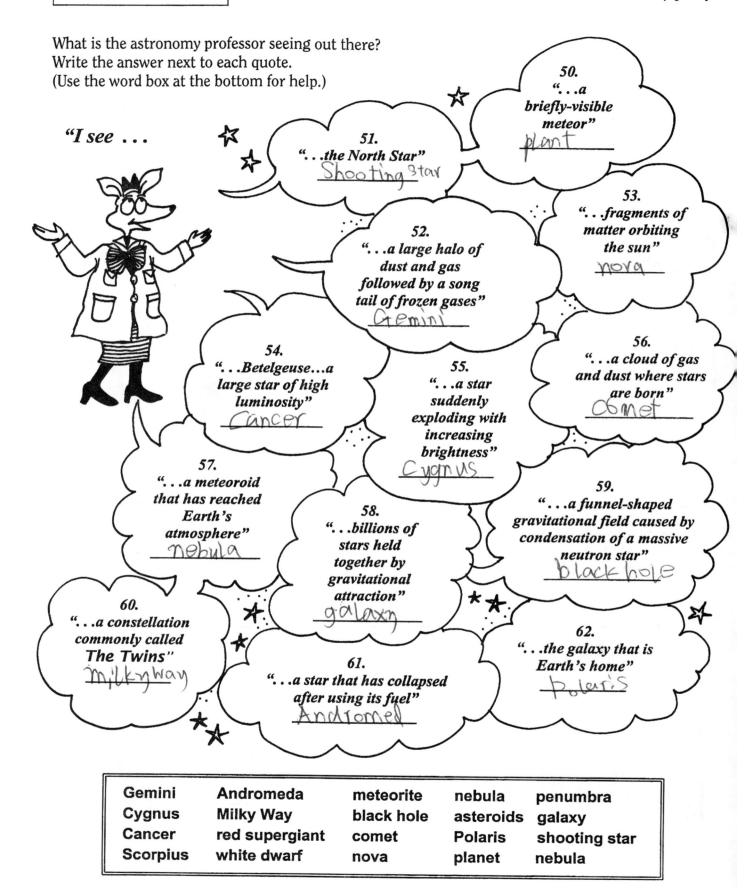

63. What do comets orbit? Earth

64. Where does a comet's tail point? Sun

69. First craft to orbit another planet:
a. Mariner 9   (b.) Apollo 11
b. Luna 2   c. Sputnik I

65. Missions that took men to the moon:
a. Gemini   c. Mariner
b. Columbia   (d.) Apollo

70. First U.S. astronaut to travel in space:
a. John Glen   c. Sally Ride
(b.) Neil Armstrong   d. Alan Shepard

66. First human walked on the moon in
a. 1957   (b.) 1969   c. 1989   d. 1993

71. First person to travel in space:
a. Neil Armstrong   c. Alan Shepard
b. Uri A. Gagarin   (d.) Michael Collins

67. First man-made object to orbit Earth
a. Sputnik   c. Luna I
(b.) Pioneer   d. Voyager

72. First woman to travel in space
a. Valentina Tereshkova   c. Shannon Lucid
(b.) Christa McAulllife   d. Sally Ride

68. First living being in space:
a. a Russian man   c. a monkey
(b.) an American man   d. a dog

73. In the moon missions, which part of the spacecraft returned to Earth?
   a. the rocket   b. the lunar module   c. the command module   (d.) the launch craft

74. Which would not be a purpose for a satellite?
   a. military operations   d. hurricane prevention
   b. communications   (e.) solar study
   c. weather forecasting   f. study of radiation around Earth

75. What is the largest telescope ever to be put into orbit?
   rocket

Earth & Space Science, Test #2

EARTH COMPOSITION

Name _____   Possible Correct Answers: 50

Date _____   Your Correct Answers: _____

1. Professor Igneous paddles a kayak between small islands in a long chain of islands. She is visiting
   a. an archipelago.
   b. a canyon.
   c. a caldera.
   d. a quarry.

2. Professor Igneous is standing on a cape. She is probably looking at
   a. an isthmus.
   b. an ocean.
   c. a desert.
   d. a plain.

3. She hikes across a high, flat area near a mountain range. She is probably on
   a. an atoll.
   b. a coastal plain.
   c. a delta.
   d. a plateau.

4. She is exploring an atoll. She is probably surrounded by
   a. a thick pine forest.
   b. a dense jungle.
   c. water.
   d. tall grasses.

5. The professor is interested in the relief of an area. She wants to learn about
   a. the variations in elevation.
   b. the kinds of vegetation.
   c. how to find water in the area.
   d. how to get rid of her stress.

6. She needs a map that describes the surface features of Earth. She should get
   a. a political map.
   b. a topographic map.
   c. a cultural map.
   d. a time zone map.

7. The professor is studying the land at 90° N latitude. Where is she? _____

8. The professor is working at this location: 10° S latitude and 30° E longitude. Where is she?
   a. South America
   b. Asia
   c. Africa
   d. North America
   e. Australia
   f. Europe

Circle the numbers of all TRUE statements below.
Use the diagram and your knowledge of Earth structure.

9. Soil is formed in the crust.

10. Earth's core is entirely liquid.

11. Most of Earth's interior is core.

12. The outermost layer is the mantle.

13. Earth's crust is about 35km thick.

14. The continental crust is thicker than the oceanic crust.

15. The mantle lies about 25km to 2900 km beneath the surface.

16. The Moho is the boundary between the mantle and the core.

17. The rocky, rigid layer is part of the mantle.

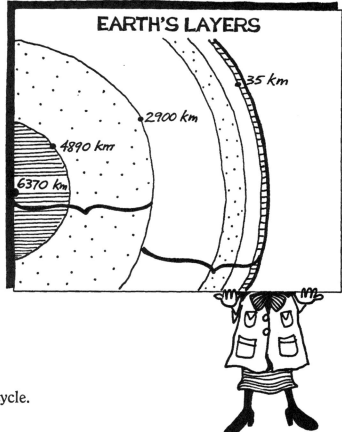

Circle the numbers of all TRUE statements.
Use the diagram and your knowledge of the rock cycle.

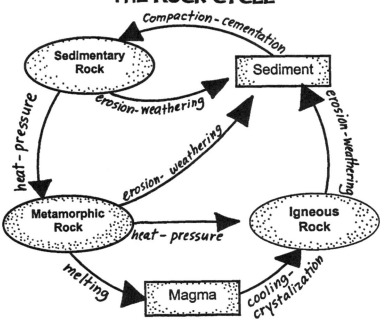

18. Heat and pressure help to form all metamorphic rocks.

19. Each kind of rock can result from another kind of rock.

20. Sedimentary rocks can only be made from metamorphic materials.

21. Metamorphic rocks can change to igneous rocks through a process of melting and cooling.

22. Weathering and erosion can break down any kind of rock.

Earth Composition

Dr. Igneous is rock-hunting this week. For each "find" described, write the number of the description in front of one of the names or characteristics shown.

She finds . . .

23. . . . a metamorphic rock formed from coal

24. . . . a porous igneous rock so light it can float

25. . . . some metamorphic rocks with a banded texture

26. . . . several rocks formed from molten magma and lava

27. . . . several rocks changed by temperature and pressure

28. . . . some ball-like objects found inside some sedimentary rocks

29. . . . some remains of once-living organisms in sedimentary rocks

30. . . . an igneous rock that is course-grained because of slow cooling

31. . . . rocks made from weathering and redeposition of loose Earth materials

32. . . . some light-colored, fine-grained igneous rocks with visible small grains

33. . . . a sedimentary rock made from round pebbles held together with cement

34. . . . a massive, nonfoliated metamorphic rock that is metamorphosed limestone

foliated
marble
pumice
conglomerate
graphite
felsites
intrusive
fossils
sedimentary
metamorphic
igneous
geodes

Write **I** *(igneous)*, **M** *(metamorphic)*, or **S** *(sedimentary)* to label the type of rock.

_____ 35. shale

_____ 36. quartzite

_____ 37. chalk

_____ 38. limestone

_____ 39. quartz

_____ 40. slate

_____ 41. gneiss

_____ 42. granite

_____ 43. marble

_____ 44. pumice

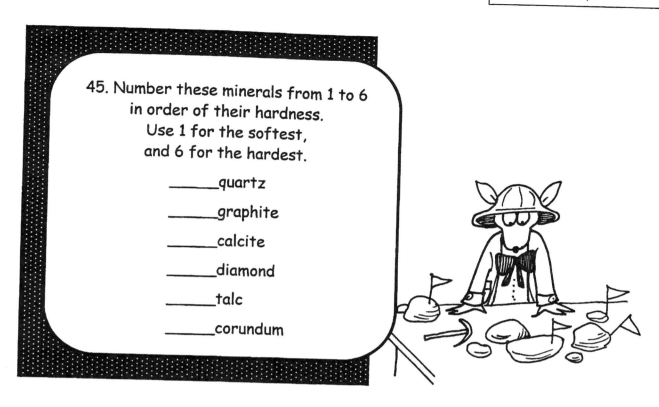

45. Number these minerals from 1 to 6 in order of their hardness. Use 1 for the softest, and 6 for the hardest.

_____ quartz
_____ graphite
_____ calcite
_____ diamond
_____ talc
_____ corundum

46. Professor Igneous hopes she has found gold. She has some shiny, yellow nuggets. The mineral cannot be scratched by a penny and leaves a greenish-black streak.

    Has she found gold?_____

47. The professor's friend has found a very soft mineral that leaves a greasy black streak on his fingers. The mineral has a shiny luster.

    What is this mineral? _____

48. The two friends have found many samples of a mineral that comes in several colors. Some of it is almost clear. It has hexagonal crystals and is hard enough to scratch a steel file.

    Could it be quartz?_____

49. Professor Igneous is holding a very soft, white mineral that flakes off. It feels soapy on her hands and leaves a powdery residue.

    She has probably found_____.

50. Next, the professor finds a deep red mineral that leaves a colorless streak. It is harder than quartz.

    Could it be garnet? _____

Earth & Space Science, Test #3

EARTH CHANGES & PROCESSES

Name _____

Date _____

Possible Correct Answers: 60

Your Correct Answers: _____

Professor Igneous and her students are headed off on an expedition to study external processes and changes on the surface of the Earth.

1. Which of these will NOT be part of the study?
   a. glaciation   b. continental drift
   c. weathering   d. erosion

2. Which is NOT an example of disintegration?
   a. A mudflow sliding down a hill uproots trees and carries away soil.
   b. In a storm, the wind picks up desert sand and deposits it miles away.
   c. Groundwater reacts with some of the elements in the rocks to form clay.

3. Which is NOT an example of decomposition?
   a. Water trickles into the cracks of a rock. The water freezes, splitting the rock apart.
   b. Rain water, a weak acid, dissolves limestone rock creating a sinkhole.
   c. River water dissolves some materials from its bed, and carries them along in solution.

4. Which is NOT an example of a way that gravity erodes land?
   a. rockfalls   b. landslides   c. creep   d. sand dunes   e. slump

5. Which of these mass movements are slow movements? (Choose one or more answers.)
   a. rockfalls   b. landslides   c. mudflows   d. creep   e. slump

6. Name the four agents of erosion.
   _____
   _____
   _____
   _____

7. Explain the difference between weathering and erosion.
   _____
   _____
   _____
   _____
   _____

Earth Changes & Processes

sand dunes
ABRASION
desert pavement
oasis
loess deposits deflation

"Oh, what the wind can do!"

Write the word or phrase that answers the question.

8. Wind is eroding Earth's surface by picking up loose ground. What is this process called?

9. Loose fragments of rock and soil are swept cleanly away from the ground leaving only the rock surface behind. What is this surface of heavy boulders and pebbles is called?

10. Land has been eroded by winds down to a depth where water is present. Vegetation grows in the desert. What has been formed?

11. The wind has gathered fine dust particles from deserts or dry riverbeds, and deposited them on hills and in valleys. What are these deposits called?

12. Particles carried by the wind scour and polish rock surfaces like a sandblasting machine. What is this process called?

13. Blowing sand grains get caught against a clump of vegetation and build up piles of sand. What are these piles called?

14. In the 1930s, the Great Plains of the U.S. suffered a drought. The winds carried away huge amounts of topsoil and ruined much land. Why didn't the vegetation hold the soil in place?

Earth Changes & Processes

A. Moving water runs rapidly off rock that has no spaces for water to soak in.	B. The running water begins to drop more sediment.	C. A moving river rolls heavy materials along the river bottom.	D. Moving water picks up soil, rocks and debris —then drops it along the way.
E. Groundwater dissolves minerals in rocks, leaving large holes.	F. Moving rivers pick up lighter sediment and carry it along as they flow.	G. A river deposits sediment in its mouth in a fan shape as it slows to enter the ocean.	H. As a moving river drags material along, it cuts a path into the rock
I. During floods, the river deposits fertile soil on either side of its banks.	J. Precipitation flows across Earth's surface and back into the ocean.	K. The running water begins to pick up more sediment.	L. Groundwater running through caves dissolves minerals, then drops them as it evaporates.

The words below have something to do with changes that running water brings to the Earth's surface. Match each word with the correct description above.
Write the correct letter.

_____ 15. runoff

_____ 16. delta

_____ 17. flood plain

_____ 18. channel

_____ 19. impervious

_____ 20. bedload

_____ 21. sinkholes & caves

_____ 22. stalactites & stalagmites

_____ 23. increased velocity

_____ 24. suspended load

_____ 25. decrease in velocity

_____ 26. sediment

What's wrong with these statements? Every statement has an error. Cross out and replace any words necessary to make each statement correct.

27. Most glaciers do not move.

28. Valley glaciers extend onto plains.

29. Glaciers make minor changes in the surfaces they cover.

30. The movement of glacial ice is called glaciation.

31. Plucking is the process of scouring the bedrock as the glacier moves over it.

32. As the velocity of the glacier decreases, the material is dropped in layers. The layered deposits of glacial debris are called tills.

33. Terminal moraines are formed from glacial debris dropped by the glacier from its base when it melts.

34. Meltwater freezes in the cracks of rocks and splits the rocks apart. The glacier picks up the broken rocks in a process called abrading, and moves the rocks along as it continues on its way.

35. The first material that a glacier drops as it melts is an unsorted jumble of rocks, clay, and sand called outwash.

36. Glaciers create V-shaped valleys.

37. Glaciers make huge scrapes in the land. These are called cirques.

38. Ground moraines are formed from glacial debris deposited in ridges at the edges of the glacier as it melts.

Earth Changes & Processes

VOLCANO

Write the letter of the diagram that shows . . .

_____ 39. a stratovolcano.

_____ 40. a dome volcano.

_____ 41. a cinder cone volcano.

_____ 42. a shield volcano.

_____ 43. Which volcano shown is the type formed by the most violent eruptions?

44. A volcano that is not active but could become active again is
    a. active. c. extinct.
    b. dormant. d. intermittent.

45. _____ is the molten rock which boils below the surface of a volcano and rises through cracks in the surface.

46. When a volcano "blows its top," it leaves a hole called a

    _____.

47. A _____ results when pressure beneath Earth's surface pushes boiling water through an opening in the surface, shooting a fountain high into the air.

These are statements about what scientists believe are causes of Earthquake activity.

Some of them are NOT correct.

Circle the numbers of the incorrect statements.

48. Earthquakes occur because molten material under pressure within the Earth forces the surface to crack.

49. Earthquakes occur because of movement within the mantle of Earth which builds up stress on the more brittle rock above it, causing cracks and shockwaves.

50. There are particular zones of earthquake activity around the Earth. One of these is around the edges of the Pacific Ocean.

51. The Earth is made of rigid plates, which float on more plastic surfaces beneath them.

52. Every once in a while, the plates move.

53. The movement of these plates explains some of the major processes of change in Earth's crust.

54. Volcanic and earthquake activity occurs most frequently in the center of these plates.

55. Faults occur where the plates move past each other.

56. Dr. Igneous is using a scale to measure the magnitude of an earthquake.

   What is this scale? _____

57. The professor happened to be standing at the point on Earth's surface where the seismic activity of an earthquake was the strongest.

   Where is she? _____

58. Last year, Professor Igneous watched a 100-foot wave that was created when a quake occurred under the ocean.

   What did she witness? _____

59. Professor Igneous lives near a fracture beneath Earth's surface, where movement takes place.

   She lives near a _____.

60. The professor spends some time working with other scientists who try to determine the age of Earth and its rocks. They use a method of measuring time based on radioactive decay. This method of measuring time in Earth history is called _____.

Earth & Space Science, Test #4

RIVERS, STREAMS, LAKES, & GROUNDWATER

Name _____

Date _____

Possible Correct Answers: 25

Your Correct Answers: _____

1. A small boat is floating in a river. It comes to a place where the river has flowed for many years across rock, eroding the softer rock. This erosion has caused
   a. an increase in the slope of the river valley.
   b. a decrease in the slope of the river.
   c. the river to narrow into a creek.

2. The boat comes to a place where the rock has been eroded so much that the rock face is vertical. The boat will probably
   a. slow down abruptly.　　　b. run into the rock.　　　c. go over a waterfall.

3. The boat floats into a river gorge. This area has probably resulted from
   a. a fast-moving river flowing down a steep slope over hard rocks.
   b. a slow-moving river flowing down a steep slope over soft rocks.
   c. a fast-moving river flowing down a gentle slope.

4. Later on, the boat floats through a wide valley that is an open V-shape. This area has probably resulted from
   a. a river that has been flowing rapidly over hard rocks.
   b. a river that has been flowing slowly over soft rocks.
   c. a fast-moving river flowing down a steep slope over hard rocks.

5. The boat floats along into the loop of a meander in the river. In this curve, where will the boat move the fastest?
   a. The boat will move faster along the inside of the curve.
   b. The boat will move faster along the outside of the curve.
   c. The speed will be the same either place.

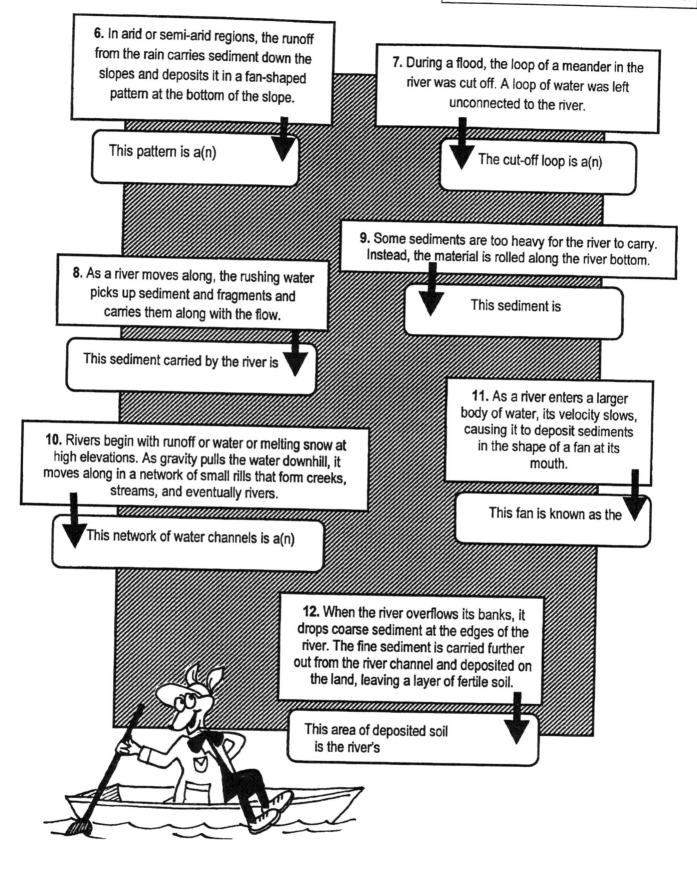

13. What drainage pattern is shown?

a. trellis
b. rectangular
c. dendritic
d. radial

14. Which is NOT a river deposit?
   a. floodplain   b. delta   c. alluvial fan   d. levee

Professor Agnes Igneous takes her students to study Lake Batholith each summer

15. For this body of water to be a lake, which of these must be true? (Circle one or more.)
   a. It must be surrounded by land.
   b. It must contain fresh water.
   c. Its bed must be lower than the lowest part of the rim.
   d. It must be fed with water at the same rate or at a greater rate than the water being lost.

16. If Lake Batholith is a reservoir, what is true about it?
   a. It was formed in the crater of a volcano.
   b. It lies along a major fault line.
   c. It was formed when a river valley was blocked by glacial debris.
   d. It was created when a dam was built across a river valley.

17. Professor Igneous takes the students on a side trip to visit a shallow lake surrounded by plants. For hundreds of years, leaves and flowers from the plants have fallen into this lake. The plants have decayed into a layer of oozing peat beneath the shallow water.

   What is it that the group is visiting? _____

18. **Which statements are true?**
(Circle one or more answers.)

A. The movement of groundwater is due to gravity.
B. The speed of groundwater movement depends only on the permeability of the rocks.
C. Groundwater moves faster than water at the surface.
D. Groundwater never returns to the surface.
E. Groundwater can only move through permeable rock.
F. Groundwater is all the water that runs across or into the ground.
G. In the zone of aeration, all the pores in the rock are filled with water.

19. Groundwater moves down into the rock until it reaches a(n) _____ rock layer.

20. _____ are permeable rock layers that are filled with water.

21. The odor in hot springs is due to _____

22. The upper surface of the zone of saturation in permeable rock is the _____.

23. Water flowing out of a crack in the surface due to pressure from the weight of a water column in an aquifer is an _____.

24. Which feature is NOT formed with the help of groundwater?
   a. cave
   b. sinkhole
   c. stalactite
   d. fumarole

25. Which of these are NOT good aquifers?
   a. clay
   b. solid rock
   c. sand
   d. gravel

Earth & Space Science, Test #5

OCEANS

Name _____

Date _____

Possible Correct Answers: 45

Your Correct Answers: ____

1. Professor Igneous found this information about Earth's oceans. Is everything on this chart true?
   yes   no

   Could it be true?
   The depth of the oceans is constantly changing.
   The oceans are giant mixtures of many substances.
   The Pacific Ocean covers about 64 million square miles.
   The size and locations of the oceans are constantly changing.
   The oceans are part of Earth's watery layer called the lithosphere.

2. Surface currents are caused by
   a. shallow water.   c. wind.
   b. temperature.   d. salinity.

3. Most surface currents north of the equator move
   a. clockwise.
   b. counterclockwise.

4. Circulation in deep water is caused by
   a. wind.   c. volcanic activity.
   b. gravity. d. density currents.

5. The density of ocean water is affected by
   a. salinity.   c. neither.
   b. temperature.   d. both.

6. Evaporation of salt water causes the density of the water to
   a. increase.   b. decrease.

7. Cold water is _____ dense than warm water.
   a. more
   b. less

8. As salinity of seawater increases, the density
   a. increases.
   b. decreases.

9. When polar water is diluted by melting ice, the water becomes
   a. more dense   b. less dense

10. When density differs, water moves from a place
    a. of greater to less density.
    b. of less to greater density.

11. What might be found in an area of upwellings?
    a. swimmers looking for warm water brought by the upwellings
    b. surfers looking for big waves created by the upwellings
    c. fishing vessels looking for fish brought to the area by the high level of nutrients

12. What is a wave?_____

13. What causes waves?_____

14. How do particles of water move in a wave?_____

15. A buoy is anchored out in a bay. The length of time it takes two consecutive waves to pass the buoy is called the wave _____.

16. A wave moving in water deeper than ½ its wavelength is a _____ wave.

17. Which of the following things happen when water moves into shallower water?
    a. Wavelength increases.
    b. Wavelength shortens.
    c. Wave height increases.
    d. Speed increases.
    e. The top of the wave moves faster than the bottom.
    f. The circulating water hits the bottom.
    g. The waves become lopsided.
    h. Wave height decreases.

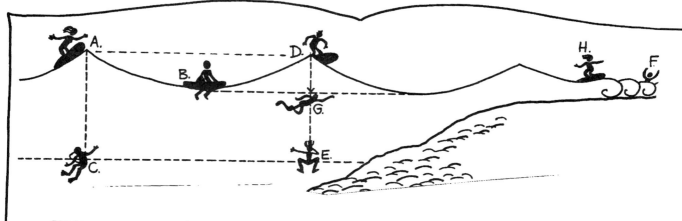

Write one or more letters for each answer.

18. Which of the swimmers, surfers, or divers are on the crest of waves?_____

19. Which are in shallow-water waves? _____

20. Which are in the troughs of waves? _____

21. Which are at a wave base? _____

22. The wavelength is the distance between _____ and _____.

23. The wave height is the distance between _____ and _____.

Oceans

Label each diagram SPRING TIDE or NEAP TIDE.

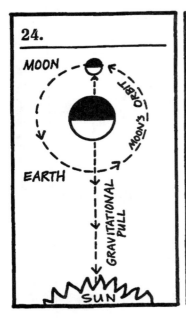
24. _____

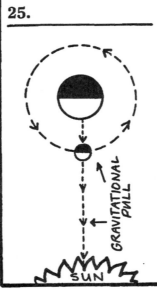
25. _____

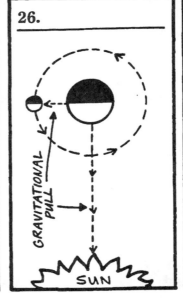
26. _____

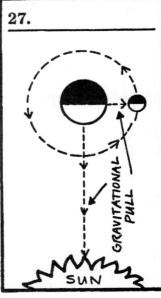
27. _____

Use one or more of these words to finish each statement.

sun   neap   moon
Earth   full moon
spring   gravity   new moon
rotational force
third quarter
first quarter

28. Tides are shallow water waves caused by gravitational attraction among _____

29. The gravitational force of the _____ causes a bulge of ocean water toward the moon.

30. The _____ of the Earth-moon system causes a bulge of ocean water on the side of Earth opposite from the moon.

31. During _____ tides, the high tides are the highest, and the low tides are the lowest.

32. The tidal range is the least during periods of _____ tides.

33. Spring tides occur when the moon is in the _____ or the _____ position.

Name _____ 146
Middle Grade Book of Science Tests   Copyright ©2001 by Incentive Publications, Inc., Nashville, TN.

34. On a beach vacation, Professor Igneous has explored the shore zone. She has found notches, stacks, and marine terraces. Which kind of shore zone is she visiting?
    a. a gently-sloping, sandy shore zone   b. a steep rocky shore zone

35. As she visits coral reefs around the world, Dr. Igneous examines the organisms that create the reef. These organisms are
    a. plants   b. benthos   c. algae   d. fish

36. Which conditions must be present for coral reefs to grow? (Circle one or more answers.)
    a. warm water   b. sediment-free water   c. deep water   d. freely-circulating water

37. Which kind of reef is shown here?
    a. barrier
    b. fringe
    c. atoll

38. When a longshore current deposits its load of sand as it enters a bay, it forms
    a. a beach.
    b. a spit.
    c. a tombolo
    d. a barrier island.

39. A shore deposit made from loose fragments that are deposited parallel to the shore is
    a. a beach.
    b. a spit.
    c. a tombolo.
    d. a barrier island.

Write the letter that shows these ocean-floor features.

_____ 40. continental shelf   _____ 43. abyssal plain

_____ 41. seamount   _____ 44. continental slope

_____ 42. oceanic trench   _____ 45. island

Earth & Space Science, Test #6

AIR, CLIMATE, & WEATHER

Name _____

Date _____

Possible Correct Answers: 45

Your Correct Answers: _____

1. Which layer contains the ozone layer?
   _____

2. What is the layer where all weather occurs?
   _____

3. What is the ceiling of the weather zone?
   _____

4. Which layer is filled with electronically charged particles?
   _____

5. Which has the coldest temperatures?
   _____

6. In which layer do the temperatures increase with increasing height?
   _____

7. The jet stream flows just below this. What is it?
   _____

8. Which layer of Earth's atmosphere contains dust, water vapor, and 75% of all gases?
   _____

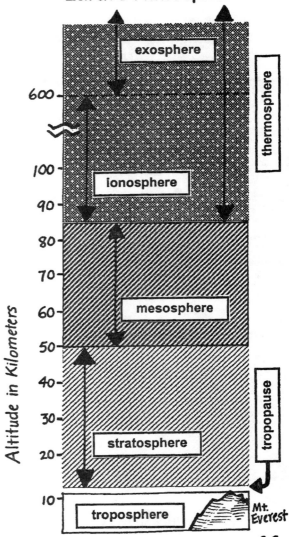

Earth's Atmosphere

9. Atmospheric pressure is greatest
   a. in the stratosphere.
   b. at the top of the troposphere.
   c. at sea level.
   d. in the ozone layer.

The professor is studying air movements all over the world. In each example, she is caught in some form of moving air. Write the word that describes her situation.

Word bank: tornado, polar easterlies, land breeze, blizzard, sea breeze, GALE, waterspout, doldrums, cyclone, HURRICANE, prevailing westerlies, trade winds

_____ 10. At nighttime, the warm air over the ocean is rising, to be replaced by the cooler air from over the land. Her plane is moving along on this.

_____ 11. She is spinning out of control in a low-pressure system where air is whirling counterclockwise.

_____ 12. Now she's flying in a windless zone along the equator.

_____ 13. She heads out over the water and is caught in a funnel of water.

_____ 14. Her plane rides along on these winds at about 20° S latitude.

_____ 15. Now, warm air moving toward the North Pole at 45°N latitude pushes her along at a great speed as she flies from southwest to northeast.

_____ 16. Warm air rises over the land, and cool air from over the sea moves in to replace it.

_____ 17. Now she is in cold, dry horizontal air currents at about 75° S latitude.

_____ 18. She is caught in the eye of a storm with warm, moist air rotating around her!

Air, Climate, & Weather

A. Get ready for a weekend of high winds and heavy, blowing snow.

B. This evening, air will expand several times at great speeds, causing booming sounds.

C. Tomorrow morning you can expect to find that the cooling ground will cool the air around it, causing the dew to freeze.

D. The temperature of the air is getting warmer at higher altitudes, holding the cold air down near the ground, causing fog in the valleys.

E. The whole area is experiencing a prolonged period without any precipitation.

F. Tomorrow, water will freeze on ice pellets in the clouds and make crystals which will join with other crystals and fall to the ground.

G. Whirling funnels of air are expected to form between the bottom of storm clouds and the ground.

H. I predict that water droplets will form and hover over the ground because the air is being cooled by the ground.

I. Centers of high pressure will bring dry, sunny, settled weather for the next several days.

J. Tomorrow morning you can expect the ground to get cold enough for water vapor in the air to condense into droplets that will be deposited on the ground.

K. Droplets of water will freeze in layers around a nucleus of ice. These will become larger as air currents toss them up and down during a severe thunderstorm.

L. I'll be right back to tell you about the measure of the amount of water vapor in the air.

M. Right now, raindrops are falling through a layer of air that is less than –3° C. This is causing the rain to freeze.

Read the weather reports above. Write the letter of one to match each weather condition.

____ 19. dew

____ 20. fog

____ 21. tornado

____ 22. snow

____ 23. highs

____ 24. hail

____ 25. frost

____ 26. thunder

____ 27. temperature inversion

____ 28. relative humidity

____ 29. sleet

____ 30. blizzard

____ 31. drought

For each example below, write the name of the front that is pictured or described. Write **WARM, COLD, STATIONARY,** or **OCCLUDED.**

_____ 32. Diagram A

_____ 33. Diagram B

_____ 34. Diagram C

_____ 35. Diagram D

_____ 36. Two cool air masses merge, pushing warm air up, bringing wind and precipitation.

_____ 37. A cold air mass invades a warm air mass, pushing warm air up steeply, bringing cooler, clearer weather.

_____ 38. A warm air mass meets a cold air mass, and slides up over it, bringing clouds and rain or snow.

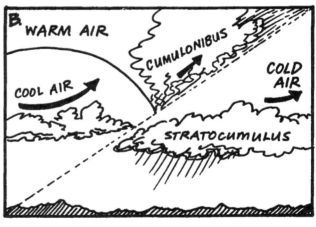

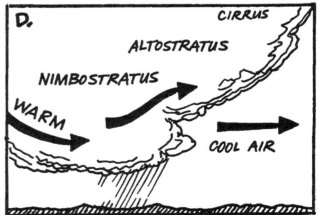

Air, Climate, & Weather

What climate type should Professor Igneous visit in each of these examples?
Write a climate zone for each location she is seeking.

39. She wants to find a place that is near the water with a mild summer and a good bit of moisture during the winter. She'd like mild winter temperatures-around 40°–50° F.

   _____

40. She wants to find a good spot where her family can ski in June.

   _____

41. In the winter, she wants to find a place that has hot, dry desert conditions.

   _____

42. For next August, the professor wants a vacation spot on a west-facing coast with a very warm dry summer and a mild, wet winter.

   _____

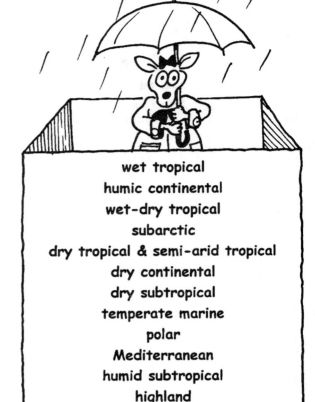

wet tropical
humic continental
wet-dry tropical
subarctic
dry tropical & semi-arid tropical
dry continental
dry subtropical
temperate marine
polar
Mediterranean
humid subtropical
highland

43. Next year, she'd like to take a safari in a place that has tall grasses, and a long, dry winter season with warm temperatures.

   _____

44. She'd like to find a place with cool temperatures, little precipitation, and nighttime frosts in July. She also wants to study permafrost while she's there.

   _____

45. She wants to take her last trip to a place where there is a rain shower every afternoon in a climate that is warm and humid all year.

   _____

KEEPING TRACK OF SKILLS

Student Progress Record 154

Class Progress Record
       (Science Concepts & Processes, Life Science) 156

Class Progress Record
       (Human Body & Health, Physical Science) 157

Class Progress Record
       (Earth & Space Science) 158

Good Skill Sharpeners for Science 159

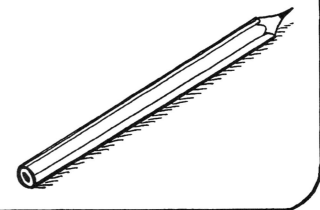

STUDENT PROGRESS RECORD – SCIENCE SKILLS

Student Name _____

TEST DATE	SCIENCE CONCEPTS & PROCESSES TESTS	SCORE ____ OF ____	COMMENTS & NEEDS
	Test # 1 Nature & History of Science	of 75	
	Test # 2 Science Concepts	of 30	
	Test # 3 Science Processes	of 40	

TEST DATE	LIFE SCIENCE TESTS TESTS	SCORE ____ OF ____	COMMENTS & NEEDS
	Test # 1 Life Characteristics & Processes	of 40	
	Test # 2 Simple Organisms	of 45	
	Test # 3 Plants	of 60	
	Test # 4 Animals	of 90	
	Test # 5 Ecology	of 35	

TEST DATE	HUMAN BODY & HEALTH TESTS TESTS	SCORE ____ OF ____	COMMENTS & NEEDS
	Test # 1 Body Systems, Part 1	of 80	
	Test # 2 Body Systems, Part 2	of 60	
	Test # 3 Diseases & Disorders	of 35	
	Test # 4 Fitness & Health	of 70	

Middle Grade Book of Science Tests — Copyright ©2001 by Incentive Publications, Inc., Nashville, TN.

STUDENT PROGRESS RECORD — SCIENCE SKILLS

Student Name _____

TEST DATE	PHYSICAL SCIENCE TESTS	SCORE ___ OF ___	COMMENTS & NEEDS
	Test # 1 Matter: Structure & Properties	of 60	
	Test # 2 Matter: Changes & Interactions	of 65	
	Test # 3 Force & Motion	of 35	
	Test # 4 Energy	of 90	
	Test # 5 Waves, Sound, & Light	of 35	

TEST DATE	EARTH & SPACE SCIENCE TESTS	SCORE ___ OF ___	COMMENTS & NEEDS
	Test # 1 Space	of 80	
	Test # 2 Earth Composition	of 50	
	Test # 3 Earth Changes & Processes	of 60	
	Test # 4 Rivers, Streams, Lakes, & Groundwater	of 25	
	Test # 5 Oceans	of 45	
	Test # 6 Air, Weather, & Climate	of 45	

CLASS PROGRESS RECORD — SCIENCE SKILLS
SCIENCE SKILLS (Science Concepts & Processes; Life Science)

Class _____ Teacher _____

SCIENCE CONCEPTS & PROCESSES TESTS

TEST DATE	TEST	COMMENTS ABOUT RESULTS	SKILLS NEEDING RE-TEACHING
	Test # 1 Nature & History of Science		
	Test # 2 Science Concepts		
	Test # 3 Science Processes		

LIFE SCIENCE TESTS

TEST DATE	TEST	COMMENTS ABOUT RESULTS	SKILLS NEEDING RE-TEACHING
	Test # 1 Life Characteristics & processes		
	Test # 2 Simple Organisms		
	Test # 3 Plants		
	Test # 4 Animals		
	Test # 5 Ecology		

CLASS PROGRESS RECORD — SCIENCE SKILLS
SCIENCE SKILLS (Human Body & Health; Physical Science)

Class _____   Teacher _____

HUMAN BODY & HEALTH TESTS

TEST DATE	TEST	COMMENTS ABOUT RESULTS	SKILLS NEEDING RE-TEACHING
	Test # 1 Body Systems, Part 1		
	Test # 2 Body Systems, Part 2		
	Test # 3 Diseases & Disorders		
	Test # 4 Fitness & Health		

PHYSICAL SCIENCE TESTS

TEST DATE	TEST	COMMENTS ABOUT RESULTS	SKILLS NEEDING RE-TEACHING
	Test # 1 Matter: Structure & Properties		
	Test # 2 Matter: Changes & Interactions		
	Test # 3 Force & Motion		
	Test # 4 Energy		
	Test # 5 Waves, Sound, & Light		

Name _____

Copyright ©2001 by Incentive Publications, Inc., Nashville, TN.

Middle Grade Book of Science Tests

CLASS PROGRESS RECORD – SCIENCE SKILLS
SCIENCE SKILLS (Earth & Space Science)

Class _____ Teacher _____

EARTH & SPACE SCIENCE TESTS

TEST DATE	TEST	COMMENTS ABOUT RESULTS	SKILLS NEEDING RE-TEACHING
	Test # 1 Space Science		
	Test # 2 Earth Composition		
	Test # 3 Earth Changes & Processes		
	Test # 4 Rivers, Streams, Lakes, & Groundwater		
	Test # 5 Oceans		
	Test # 6 Air, Weather, & Climate		

GOOD SKILL SHARPENERS FOR SCIENCE

The tests in this book will identify student needs for practice, re-teaching or reinforcement of basic science skills. Once those areas of need are known, then what? You and your students need to find some good ways to strengthen those skills.

The BASIC/Not Boring Skills Series, published by Incentive Publications (www.incentivepublications.com), offers 20 books to sharpen basic skills at the Grades 6–8 level. Three of these books are full of science exercises. The books cover these topics: Life Science, Physical Science, and Earth & Space Science.

The pages of these books are student-friendly, clever, and challenging—guaranteed not to be boring! They cover a wide range of skills, including the skills assessed in this book of tests. A complete checklist of skills is available at the front of each book, complete with a reference list directing you to the precise pages for polishing those skills.

TEST IN THIS BOOK **Middle Grade Book of Science Tests**	Pages in this Book	You will find pages to sharpen skills in these locations from the BASIC/Not Boring Skills Series, published by Incentive Publications.
Science Concepts & Processes Test # 1 **History & Nature of Science**	12–19	Gr. 6–8 Life Science Gr. 6–8 Physical Science Gr. 6–8 Earth & Space Science
Science Concepts & Processes Test # 2 **Science Concepts**	20–25	Gr. 6–8 Life Science Gr. 6–8 Physical Science Gr. 6–8 Earth & Space Science
Science Concepts & Processes Test # 3 **Science Processes**	26–31	Gr. 6–8 Life Science Gr. 6–8 Physical Science Gr. 6–8 Earth & Space Science
Life Science Test # 1 **Life Characteristics & Processes**	35–39	Gr. 6–8 Life Science
Life Science Test # 2 **Simple Organisms**	40–43	Gr. 6–8 Life Science
Life Science Test # 3 **Plants**	44–51	Gr. 6–8 Life Science
Life Science Test # 4 **Animals**	52–59	Gr. 6–8 Life Science
Life Science Test # 5 **Ecology**	60–65	Gr. 6–8 Life Science

(continued on next page)

Name _____

TEST IN THIS BOOK Middle Grade Book of Science Tests	Pages in this Book	You will find pages to sharpen skills in these locations from the BASIC/Not Boring Skills Series, published by Incentive Publications.
Human Body & Health Test # 1 **Body Systems, Part 1**	69–75	
Human Body & Health Test # 2 **Body Systems, Part 2**	76–81	
Human Body & Health Test # 3 **Diseases & Disorders**	82–85	
Human Body & Health Test # 4 **Fitness & Health**	86–91	
Physical Science Test # 1 **Matter: Structure & Properties**	94–101	Gr. 6–8 Physical Science
Physical Science Test # 2 **Matter: Changes & Interactions**	102–105	Gr. 6–8 Physical Science
Physical Science Test # 3 **Force & Motion**	106–109	Gr. 6–8 Physical Science
Physical Science Test # 4 **Energy**	110–117	Gr. 6–8 Physical Science
Physical Science Test # 5 **Waves, Sound, & Light**	118–121	Gr. 6–8 Physical Science
Earth & Space Science Test # 1 **Space Science**	125–129	Gr. 6–8 Earth & Space Science
Earth & Space Science Test # 2 **Earth Composition**	130–133	Gr. 6–8 Earth & Space Science
Earth & Space Science Test # 3 **Earth Changes & Processes**	134–139	Gr. 6–8 Earth & Space Science
Earth & Space Science Test # 4 **Rivers, Streams, Lakes, & Groundwater**	140–143	Gr. 6–8 Earth & Space Science
Earth & Space Science Test # 5 **Oceans**	144–147	Gr. 6–8 Earth & Space Science
Earth & Space Science Test # 6 **Air, Weather, & Climate**	148–152	Gr. 6–8 Earth & Space Science

EARTH & SPACE SCIENCE ANSWER KEYS

Pages 144–147: Earth & Space Science, Test # 5: OCEANS

1. no	15. period	31. spring
2. c	16. deep water	32. neap
3. a	17. b, c, e, f, g	33. new and full
4. d	18. A, D	34. b
5. d	19. H, F	35. b
6. a	20. B	36. a, c, d
7. a	21. C, E	37. a
8. a	22. A, D	38. b
9. b	23. D, G	39. a
10. a	24. Spring Tide	40. S
11. c	25. Spring Tide	41. X
12. movements where water rises and falls	26. Neap Tide	42. W
	27. Neap Tide	43. U
13. winds, tides, or earthquakes	28. sun-moon-Earth	44. T
	29. moon	45. Y
14. in a circle	30. rotational force	

Pages 148–152: Earth & Space Science, Test # 6: AIR, WEATHER, & CLIMATE

1. stratosphere	17. polar easterlies	33. occluded
2. troposphere	18. hurricane	34. cold
3. tropopause	19. J	35. warm
4. ionosphere	20. H	36. cold
5. mesosphere	21. G	37. occluded
6. thermosphere	22. F	38. warm
7. tropopause	23. I	39. temperate marine
8. troposphere	24. K	40. highland or subarctic
9. c	25. C	41. dry tropical or semi-arid tropical
10. land breeze	26. B	
11. cyclone	27. D	42. Mediterranean
12. doldrums	28. L	43. wet-dry tropical
13. waterspout	29. M	44. subarctic
14. trade winds	30. A	45. wet tropical
15. prevailing westerlies	31. E	
16. sea breeze	32. stationary	